MICROSCOPES
BEST UTILIZATION

MICROSCOPES

BEST UTILIZATION

By

SANANDA CHATTERJEE

DAE, DCA, CMCNet

SCIENTFIC CONSULTANT

INSTITUTE FOR NATURAL SCIENCES

KOLKATA

DISCOVERY PUBLISHING HOUSE PVT. LTD.

NEW DELHI-110 002

Published by:
Tilak Wasan
DISCOVERY PUBLISHING HOUSE PVT. LTD.
4383/4B, Ansari Road, Darya Ganj
New Delhi-110 002 (India)
Phone : +91-11-23279245; 23253475; 43596065
E-mail: discoverybooksindia@gmail.com
discoverypublishinghouse@gmail.com
namitwasan9@gmail.com
web : www.discoverypublishinggroup.com

Reprinted: 2019

First Edition: 2010

ISBN: 978-81-8356-639-1

Microscopes – *Best Utilization*

Printed at:
Infinity Imaging Systems
Delhi

Preface

Microscopes' best utilization is a small part of the vast subject. Instrumentation technology, in this book we will see the different microscopes and their uses, starting from the basic compound microscopes to the most advanced confocal microscopes. Their different parts and the methods to use them properly have also been explained in this book within 8 chapters.

Right selection of the microscopes will give a person the appropriate vision in the world of microbes, rather the "very little living animalcules" of Sir Antony van Leeuwenhoek. It must not be forgotten in any circumstances that without the discovery of this man our modern medicine would have been in a complete dark.

This book is, therefore, a heartiest tribute to that gallantry figure by the grace of whom many subjects have come out and has given our modern medical science a sincere direction towards a fruitful success.

Here in this book as it will be seen about different microscopes with their own working principles are therefore manufactured by different companies.

This should be kept in mind that always a standard instrument is available against standard price, but when a sudden price drop is noticed, it should be noted well that somewhere in some level and because of some reason a compromise is being done.

Regarding the preparation of the manuscript a thought was continuously prevailing in mind about a famous saying "Pen Paper and Mind, writes confined". The emphasis on the microscopes only was not at all price worthy because a bad slide or wrongly prepared specimen can destroy the whole analysis or study. So an equimportance has been given to the preparation of the slides and use of stains in the book.

This book is made for those students and also for those users of microscopes who can and will certainly able to extract the essence of microscope.

This book is the small endeavour and a humble tribute to that great stallward legend Sir Antony van Leeuwenhoek, that ordinary Dutch weaver because of whom Microbiology, Bacteriology, Biochemistry has become subjects and are guiding the medical science towards a profound future.

Sananda Chatterjee

Contents

Preface

1. Introduction 1
2. Type of Microscopes 4
3. The Lenses of Microscopes 8
4. The Compound Microscopes 43
5. The Phase Contrast Microscopy 54
6. The Dark Field Microscopy 61
7. The Fluorescent and Confocal Microscopy 71
8. The Application of Microscopes 167

Antony van Leeuwenhoek

. . my work, which I've done for a long time, was not pursued in order to gain the praise. I now enjoy, but chiefly from a craving after knowledge, which I notice resides in me more than in most other men. And therewithal, whenever, I found out anything remarkable, I have thought it my duty to put down my discovery on paper, so that all ingenious people might be informed thereof.

—Antony van Leeuwenhoek. Letter of June 12, 1716

Antony van Leeuwenhoek was an unlikely scientist. A tradesman of Delft, Holland, he came from a family of tradesmen, had no fortune, received no higher education or university degrees, and knew no languages other than his native Dutch. This would have been enough to exclude him from the scientific community of his time completely. Yet with skill, diligence, an endless curiosity, and an open mind free of the scientific dogma of his day, Leeuwenhoek succeeded in making some of the most important discoveries in the history of biology. It was he who discovered bacteria, free-living and parasitic microscopic protists, sperm cells, blood cells, microscopic nematodes and rotifers, and much more. His researches, which were widely circulated, opened up an entire world of microscopic life to the awareness of scientists.

Leeuwenhoek was born in Delft on October 24, 1632. (His last name, incidentally, often is quite troublesome to non-Dutch speakers: "layu-wen-hook" is a passable English approximation.) His father was a basket-maker, while his mother's family were brewers. Antony was educated as a child in a school in the town of Warmond, then lived with his uncle at Benthuizen; in 1648 he was apprenticed in a linen-draper's shop. Around 1654 he returned to Delft, where he spent the rest of his life. He set himself up in business as a draper (a fabric merchant); he is also known to have worked as a surveyor, a wine assayer, and as a minor city official. In 1676 he served as the trustee of the estate of the deceased and bankrupt Jan Vermeer, the famous painter, who had had been born in the same year as Leeuwenhoek and is thought to have been a friend of his. And at some time before 1668, Antony van Leeuwenhoek learned to grind lenses, made simple microscopes, and began observing with them. He seems to have been inspired to take up microscopy by having seen a copy of Robert Hooke's illustrated book *Micrographia*, which depicted Hooke's own observations with the microscope and was very popular.

Leeuwenhoek is known to have made over 500 "microscopes", of which fewer than ten have survived to the present day. In basic design, probably all of Leeuwenhoek's instruments — certainly all the ones that are known — were simply powerful magnifying glasses, not compound microscopes of the type used today. A drawing of one of Leeuwenhoek's "microscopes" is shown at the left. Compared to modern microscopes, it is an extremely simple device, using only one lens, mounted in a tiny hole in the brass plate that makes up the body of the instrument. The specimen was mounted on the sharp point that sticks up in front of the lens, and its position and focus could be adjusted by turning the two screws. The entire instrument was only 3-4 inches long, and had to be held up close to the eye; it required good lighting and great patience to use.

Compound microscopes (that is, microscopes using more than one lens) had been invented around 1595, nearly 40 years before Leeuwenhoek was born. Several of Leeuwenhoek's predecessors and contemporaries, notably Robert Hooke in England and Jan Swammerdam in the Netherlands, had built compound microscopes and were making important discoveries with them. These were much more similar to the microscopes in use today. Thus, although Leeuwenhoek is sometimes called "the inventor of the microscope", he was no such thing.

However, because of various technical difficulties in building them, early compound microscopes were not practical for magnifying objects more than about 20 or 30 times natural size. Leeuwenhoek's skill at grinding lenses, together with his naturally acute eyesight and great care in adjusting the lighting where he worked, enabled him to build microscopes that magnified over 200 times, with clearer and brighter images than any of his colleagues could achieve. What further distinguished him was his curiosity to observe almost anything that could be placed under his lenses, and his care in describing what he saw. Although he himself could not draw well, he hired an illustrator to prepare drawings of the things he saw, to accompany his written descriptions. Most of his descriptions of microorganisms are instantly recognizable.

In 1673, Leeuwenhoek began writing letters to the newly-formed Royal Society of London, describing what he had seen with his microscopes — his first letter contained some observations on the stings of bees. For the next 50 years he corresponded with the Royal Society; his letters, written in Dutch, were translated into English or Latin and printed in the *Philosophical Transactions of the Royal Society*, and often reprinted separately. To give some of the flavor of his discoveries, we present extracts from his observations, together with modern pictures of the organisms that Leeuwenhoek saw.

In a letter of September 7, 1674, Leeuwenhoek described observations on lake water, including an excellent description of the green charophyte alga *Spirogyra*:

> "Passing just lately over this lake, . . . and examining this water next day, I found floating therein divers earthy particles, and some green streaks, spirally wound serpent-wise, and orderly arranged, after the manner of the copper or tin worms, which distillers use to cool their liquors as they distil over. The whole circumference of each of these streaks was about the thickness of a hair of one's head. . . all consisted of very small green globules joined together: and there were very many small green globules as well."

A letter dated December 25, 1702, gives descriptions of many protists, including this ciliate, *Vorticella*:

> "In structure these little animals were fashioned like a bell, and at the round opening they made such a stir, that the particles in the water thereabout were set in motion thereby. . . And though I must have seen quite 20 of these little animals on their long tails alongside one another very gently moving, with outstretched bodies and straightened-out tails; yet in an instant, as it were, they pulled their bodies and their tails together, and no sooner had

> they contracted their bodies and tails, than they began to stick their tails out again very leisurely, and stayed thus some time continuing their gentle motion: which sight I found mightily diverting."

On September 17, 1683, Leeuwenhoek wrote to the Royal Society about his observations on the plaque between his own teeth, "a little white matter, which is as thick as if 'twere batter." He repeated these observations on two ladies (probably his own wife and daughter), and on two old men who had never cleaned their teeth in their lives. Looking at these samples with his microscope, Leeuwenhoek reported how in his own mouth:

> "I then most always saw, with great wonder, that in the said matter there were many very little living animalcules, very prettily a-moving. The biggest sort . . . had a very strong and swift motion, and shot through the water (or spittle) like a pike does through the water. The second sort. . . oft-times spun round like a top. . . and these were far more in number."

In the mouth of one of the old men, Leeuwenhoek found

> "an unbelievably great company of living animalcules, a-swimming more nimbly than any I had ever seen up to this time. The biggest sort. . . bent their body into curves in going forwards. . .. Moreover, the other animalcules were in such enormous numbers, that all the water. . . seemed to be alive."

These were among the first observations on living bacteria ever recorded.

Leeuwenhoek looked at animal and plant tissues, at mineral crystals and at fossils. He was the first to see microscopic foraminifera, which he described as "little cockles. . . no bigger than a coarse sand-grain." He discovered blood cells, and was the first to see living sperm cells of animals. He discovered microscopic animals such as nematodes and rotifers. The list of his discoveries goes on and on. Leeuwenhoek soon became famous as his letters were published and translated. In 1680 he was elected a full member of the Royal Society, joining Robert Hooke, Henry Oldenburg, Robert Boyle, Christopher Wren, and other scientific luminaries of his day — although he never attended a meeting. In 1698 he demonstrated circulation in the capillaries of an eel to Tsar Peter the Great of Russia, and he continued to receive visitors curious to see the strange things he was describing. He continued his observations until the last days of his life. After his death on August 30, 1723, the pastor of the New Church at Delft wrote to the Royal Society:

> . . . *Antony van Leeuwenhoek considered that what is true in natural philosophy can be most fruitfully investigated by the experimental method, supported by the evidence of the senses; for which reason, by diligence and tireless labour he made with his own hand certain most excellent lenses, with the aid of which he discovered many secrets of Nature, now famous throughout the whole philosophical World.*

Chapter 1

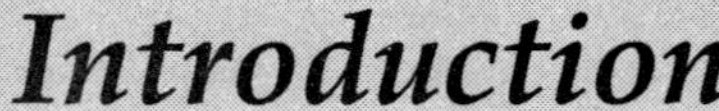

Introduction

During the 1st century AD (year 100), glass had been invented and the Romans were looking through the glass and testing it. They experimented with different shapes of clear glass and one of their samples was thick in the middle and thin on the edges. They discovered that if you held one of these "lenses" over an object, the object would look larger.

Someone also discovered that you can focus the rays of the sun with one of these special "glasses" and start a fire. These early lenses were called magnifiers or burning glasses. The word lens by the way, is derived from the Latin word lentil, as they were named because they resembled the shape of a lentil bean (look up lens in a dictionary).

These lenses were not used much until the end of the 13th century when spectacle makers were producing lenses to be worn as glasses.

The early simple "microscopes" which were really only magnifying glasses had one power, usually about 6X - 10X . One thing that was very common and interesting to look at was fleas and other tiny insects. These early magnifiers were hence called "flea glasses".

Sometime about the year 1590, two Dutch spectacle makers, Zaccharias Janssen and his father Hans started experimenting with these lenses. They put several lenses in a tube and made a very important discovery. The object near the end of the tube appeared to be greatly enlarged, much larger than any simple magnifying glass could achieve by itself! They had just invented the compound microscope (which is a microscope that uses two or more lenses).

Galileo heard of their experiments and started experimenting on his own. He described the principles of lenses and light rays and improved both the microscope and telescope. He added a focusing device to his microscope and of course went on to explore the heavens with his telescopes.

Antony Leeuwenhoek of Holland became very interested in lenses while working with magnifying glasses in a dry goods store. He used the magnifying glass to count threads in woven cloth. He became so interested that he learned how to make lenses. By grinding and polishing, he was able to make small lenses with great curvatures. These rounder lenses produced greater magnification, and his microscopes were able to magnify up to 270X!

Antony Leeuwenhoek became more involved in science and with his new improved microscope was able to see things that no man had ever seen before. He saw bacteria, yeast, blood cells and many tiny animals swimming about in a drop of water. From his great contributions, many discoveries and research papers, Antony Leeuwenhoek (1632-1723) has since been called the "Father of Microscopy".

Robert Hooke, an Englishman (who is sometimes called the "English Father of Microscopy"), also spent much of his life working with microscopes and improved their design and capabilities.

Little was done to improve the microscope until the middle of the 19th century when great strides were made and quality instruments like today's microscope emerged. Companies in Germany like Zeiss and an American company founded by Charles Spencer began producing fine optical instruments.

Today, there are no microscope manufacturers in the US and most of the microscopes come from Germany, Japan and China. Toy plastic microscopes should be avoided as they do not achieve the level of quality of the basic instruments with metal frames and glass lenses.

Because of foreign production, quality microscopes have become affordable for all. Zaccharias Janssen, the inventor of the microscope would marvel at the quality of even the most basic microscopes found in schools today.

During that historic period known as the Renaissance, after the middle Ages, there occurred the inventions of printing, gunpowder and the mariner's compass, followed by the discovery of America. Equally remarkable was the invention of the light microscope: an instrument that enables the human eye, by means of a lens or combinations of lenses, to observe enlarged images of tiny objects. It made visible the fascinating details of worlds within worlds.

Invention of Glass Lenses

Long before, in the hazy unrecorded past, someone picked up a piece of transparent crystal thicker in the middle than at the edges, looked through it, and discovered that it made things look larger. Someone also found that such a crystal would focus the sun's rays and set fire to a piece of parchment or cloth. Magnifiers and "burning glasses" or "magnifying glasses" are mentioned in the writings of Seneca and Pliny the Elder, Roman philosophers during the first century A. D., but apparently they were not used much until the invention of spectacles, toward the end of the 13th century. They were named lenses because they are shaped like the seeds of a lentil.

The earliest simple microscope was merely a tube with a plate for the object at one end and, at the other, a lens which gave a magnification less than ten diameters — ten times the actual size. These excited general wonder when used to view fleas or tiny creeping things and so were dubbed "flea glasses."

Birth of the Light Microscope

About 1590, two Dutch spectacle makers, Zaccharias Janssen and his son Hans, while experimenting with several lenses in a tube, discovered that nearby objects appeared greatly enlarged. That was the forerunner of the compound microscope and of the telescope. In 1609, Galileo, father of modern physics and astronomy, heard of these early experiments, worked out the principles of lenses, and made a much better instrument with a focusing device.

Antony van Leeuwenhoek (1632-1723)

The father of microscopy, Anton van Leeuwenhoek of Holland, started as an apprentice in a dry goods store where magnifying glasses were used to count the threads in cloth. He taught himself new methods for grinding and polishing tiny lenses of great curvature which gave magnifications up to 270 diameters, the finest known at that time. These led to the building of his microscopes and the biological discoveries for which he is famous. He was the first to see and describe bacteria, yeast plants, the teeming life in a drop of water, and the circulation of blood corpuscles in capillaries. During a long life he used his lenses to make pioneer studies on an extraordinary variety of things, both living and non living, and reported his findings in over a hundred letters to the Royal Society of England and the French Academy.

Robert Hooke

Robert Hooke, the English father of microscopy, re-confirmed Anton van Leeuwenhoek's discoveries of the existence of tiny living organisms in a drop of water. Hooke made a copy of Leeuwenhoek's light microscope and then improved upon his design.

Charles A. Spencer

Later, few major improvements were made until the middle of the 19th century. Then several European countries began to manufacture fine optical equipment but none finer than the marvelous instruments built by the American, Charles A. Spencer, and the industry he founded. Present day instruments, changed but little, give magnifications up to 1250 diameters with ordinary light and up to 5000 with blue light.

Beyond the Light Microscope

A light microscope, even one with perfect lenses and perfect illumination, simply cannot be used to distinguish objects that are smaller than half the wavelength of light. White light has an average wavelength of 0.55 micrometers, half of which is 0.275 micrometers. (One micrometer is a thousandth of a millimeter, and there are about 25,000 micrometers to an inch. Micrometers are also called microns.) Any two lines that are closer together than 0.275 micrometers will be seen as a single line, and any object with a diameter smaller than 0.275 micrometers will be invisible or, at best, show up as a blur. To see tiny particles under a microscope, scientists must bypass light altogether and use a different sort of "illumination", one with a shorter wavelength.

Chapter 2 *Type of Microscopes*

Microscopes can largely be separated into three classes: optical theory microscopes (Light microscope), electron microscopes (e.g., TEM), and scanning probe microscopes (SPM). Optical microscopes are microscopes which function through the optical theory of lenses in order to magnify the image generated by the passage of a wave through the sample, or reflected by the sample. The waves used are either electromagnetic (in optical microscopes) or electron beams (in electron microscopes). The types are the Compound Light, Stereo, and the electron microscope. Here in this book we are going to discuss only about the optical microscopes)

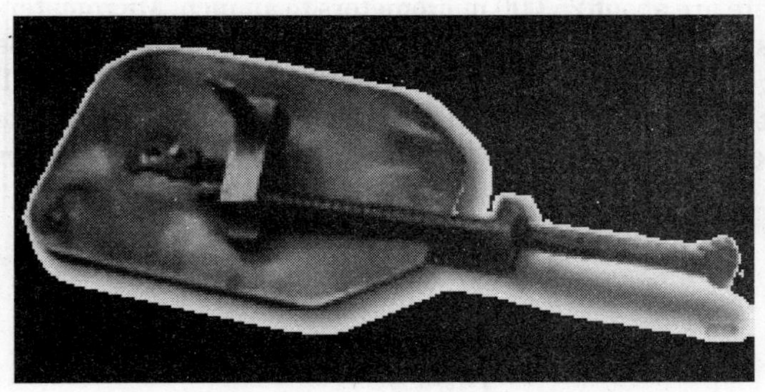

Fig. 2.1.

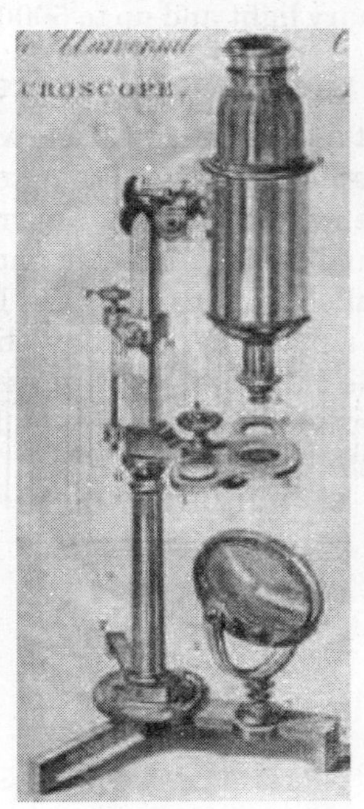

Fig. 2.1A.

Fig. 2.1 and 2.1A. The Microscope Redesigned by Robert Hooke after the invention of Anton Van Leeuwenhoek microscope.

Figure 2.2 below shows the types of microscopes properly:

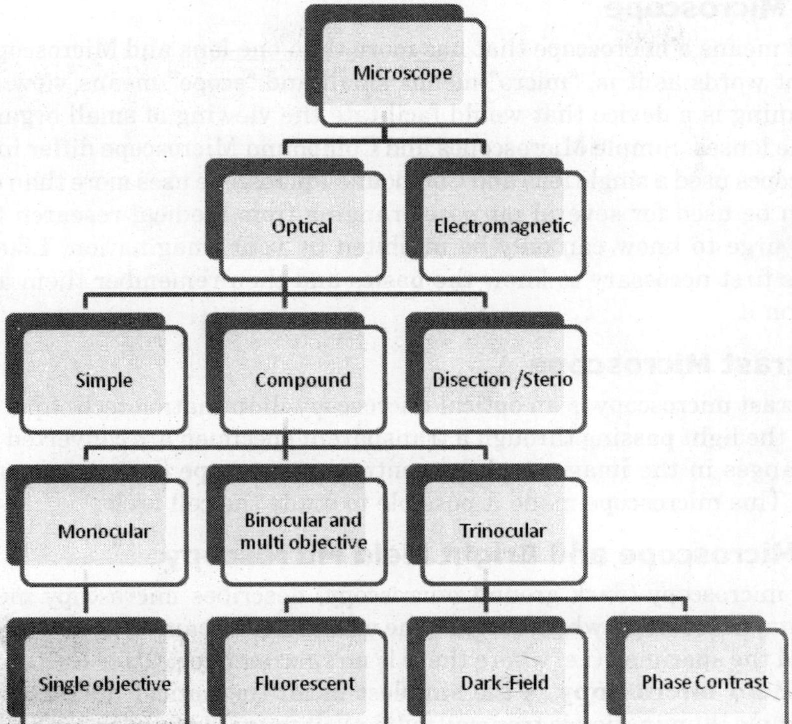

Fig. 2.2. Types of Microscopes.

Optical Microscopes

Optical microscopes, through their use of visible wavelengths of light, are the simplest and hence most widely used type of microscope. Optical microscopes typically use refractive glass and occasionally of plastic or quartz, to focus light into the eye or another light detector. Mirror-based optical microscopes operate in the same manner. Typical magnification of a light microscope, assuming visible range light, is up to 1500x with a theoretical resolution limit of around 0.2 micrometres or 200 nanometers.. Using shorter wavelengths of light, such as the ultraviolet, is one way to improve the spatial resolution of the microscope as are techniques such as Near-field scanning optical microscope.Various wavelengths of light, including those beyond the visible range, are sometimes used for special purposes. Ultraviolet light is used to enable the resolution of smaller features as well as to image samples that are transparent to the eye. Near infrared light is used to image circuitry embedded in bonded silicon devices as silicon is transparent in this region.

Scanning Electron Microscope - allow scientists to view a universe too small to be seen with a light microscope. SEMs donot use light waves; they use electrons (negatively charged electrical particles) to magnify objects up to two million times.

Transmission Electron Microscope - also uses electrons, but instead of scanning the surface (as with SEM's) electrons are passed through very thin specimens.

Stereoscope - this microscope allows for binocular (two eyes) viewing of larger specimens.

Compound Microscope

Compound means a microscope that has more than one lens and Microscope, if divided in to two different words as it is, "micro" means small and "scope" means view. Therefore in a whole the meaning is a device that would facilitate the viewing of small organisms with the help of multiple lenses. Simple Microscopes and Compound Microscope differ in the sense that Simple microscopes used a single lens and Compound Microscope uses more than one. Compound microscope can be used for several purposes ranging from medical research to a day at the beach as your urge to know can only be inhibited by your imagination. Like every field of knowledge it is first necessary to know the basics and then remember them as and how you keep working on it.

Phase Contrast Microscope

Phase contrast microscopy is an optical microscopy illumination technique in which small phase shifts in the light passing through a transparent specimen are converted into amplitude or contrast changes in the image. A phase contrast microscope does not require staining to view the slide. This microscope made it possible to study the cell cycle.

Dark Field Microscope and Bright Field Microscopy

Dark field microscopy (dark ground microscopy) describes microscopy methods, in both light and electron microscopy, which exclude the unscattered beam from the image. As a result, the field around the specimen (i.e. where there is no specimen to scatter the beam) is generally dark. **Bright field microscopy** is the simplest of all the optical microscopy illumination techniques. Sample illumination is transmitted (i.e., illuminated from below and observed from above) white light. The most common use of the microscope involves the use of an organism mounted to a glass microscope slid

Fluorescence Microscope

A fluorescence microscope (colloquially synonymous with *epifluorescent microscope*) is a light microscope used to study properties of organic or inorganic substances using the phenomena of fluorescence and phosphorescence instead of, or in addition to, reflection and absorption.

Magnification

Your microscope has three magnifications generally: Scanning, Low and High. Each objective will have written the magnification. In addition to this, the ocular lens (eyepiece) has a magnification. The total resultant magnification is the ocular multiplied by the objective.

	Magnification	*Ocular lens*	*Total Magnification*
Scanning	4x	10x	40x
Low Power	10x	10x	100x
High Power	40x	10x	400x
Oil Immersion	100x	10x	1000x

Now we shall see the different ranges of the microscope and their magnification power.

An optical microscope may have plastic or glass lenses. In its simplest form, it is a tube with one lens at one end (the eyepiece), and another lens at the other end (objective). The combination of these two lenses forms the magnification power of the microscope. So if the tube has a 5x eyepiece and a 20x objective, you will be able to observe microscopical sized objects and creatures at 100x magnification.

Optical Microscopes use natural light by way of reflected light from a mirror, or artificial light by way of a light bulb or light emitting diode to illuminate the subject being viewed. This creates a limitation in the ability of a microscope to magnify beyond a certain limit. With very well made glass lenses, the effective magnification range of all optical microscopes is approximately 1400x - 1600x. Effectiveness is about resolving detail in the subject observed with a microscope and not just about making something appear larger.

Table 2.1 shows the magnification ranges of different optical microscopes

Table 2.1. The Magnification Ranges of Different optical Microscope

Types of Microscopes	*Magnification Range*
Compound Microscopes	5X – 1000X
Phase Contrast Microscopes	4X - 1600X
Dark Field Microscopes	4X – 1800X
Fluorescent Microscopes	4X – 1800X

Chapter 3

The Lenses of Microscopes

The microscope uses two types of lenses one is the **Eye piece** and the other one is **Objective.**

Microscope Eyepieces

An **eyepiece**, or **ocular lens**, is a type of lens that is attached to a variety of optical devices such as telescopes and microscopes. It is so named because it is usually the lens that is closest to the eye when someone looks through the device. The objective lens or mirror collects light and brings it to focus creating an image. The eyepiece is placed at the focal point of the objective to magnify this image. The amount of magnification depends on the focal length of the eyepiece.

An eyepiece consists of several "lens elements" in a housing, with a "barrel" on one end. The barrel is shaped to fit in a special opening of the instrument to which it is attached. The image can be focused by moving the eyepiece nearer and further from the objective. Most instruments have a focusing mechanism to allow movement of the shaft in which the eyepiece is mounted, without needing to manipulate the eyepiece directly.

The eyepieces of binoculars are usually permanently mounted in the binoculars, causing them to have a pre-determined magnification and field of view. With telescopes and microscopes, however, eyepieces are usually interchangeable. By switching the eyepiece, the user can adjust what is viewed. For instance, eyepieces will often be interchanged to increase or decrease the magnification of a telescope. Eyepieces also offer varying fields of view, and differing degrees of eye relief for the person who looks through them. Modern research-grade telescopes do not use eyepieces. Instead, they have high-quality CCD sensors mounted at the focal point, and the images are viewed on a computer screen. Some amateur astronomers use their telescopes the same way, but direct optical viewing with eyepieces is still very common.

Eyepiece Properties

Several properties of an eyepiece are likely to be of interest to a user of an optical instrument, when comparing eyepieces and deciding which eyepiece suits their needs.

Design Distance to Entrance Pupil

Eyepieces are optical systems where the entrance pupil is invariably located outside of the system. They must be designed for optimal performance for a specific distance to this entrance pupil (i.e. with minimum aberrations for this distance). In a refracting astronomical telescope the entrance pupil is identical with the objective. This may be several feet distant from the eyepiece; whereas with a microscope eyepiece the entrance pupil is close to the back focal plane of the objective, mere inches from the eyepiece. Microscope eyepieces may be corrected differently from telescope eyepieces; however, most are also suitable for telescope use.

Elements and Groups

Elements are the individual lenses, which may come as simple lenses or "singlets" and cemented doublets or (rarely) triplets. When lenses are cemented together in pairs or triples, the combined elements are called *groups* (of lenses).

The first eyepieces had only a single lens element, which delivered highly distorted images. Two and three-element designs were invented soon after, and quickly became standard due to the improved image quality. Today, engineers assisted by computer-aided drafting software have designed eyepieces with seven or eight elements that deliver exceptionally large, sharp views.

Internal Reflection and Scatter

Internal reflections, sometimes called *scatter*, cause the light passing through an eyepiece to disperse and reduce the contrast of the image projected by the eyepiece. When the effect is particularly bad, "ghost images" are seen, called *ghosting*. For many years, simple eyepiece designs with a minimum number of internal air-to-glass surfaces were preferred to avoid this problem.

One solution to scatter is to use thin film coatings over the surface of the element. These thin coatings are only one or two wavelengths deep, and work to reduce reflections and scattering by changing the refraction of the light passing through the element. Some coatings may also absorb light that is not being passed through the lens in a process called total internal reflection where the light incident on the film is at a shallow angle.

Chromatic Aberration

Lateral chromatic aberration is caused because the refraction at glass surfaces differs for light of different wavelengths. Blue light, seen through an eyepiece element, will not focus to the same plane as red light. The effect can create a ring of false colour around point sources of light and results in a general blurriness to the image.

One solution is to reduce the aberration by using multiple elements of different types of glass. Achromats are lens groups that bring two different wavelengths of light to the same focus and exhibit greatly reduced false colour. Low dispersion glass may also be used to reduce chromatic aberration.

Longitudinal chromatic aberration is a pronounced effect of optical telescope objectives, because the focal lengths are so long. Microscopes, whose focal lengths are generally shorter, do not tend to suffer from this effect.

Focal Length

The focal length of an eyepiece is the distance from the principal plane of the eyepiece where parallel rays of light converges to a single point. When in use, the focal length of an eyepiece, combined with the focal length of the telescope or microscope objective, to which it is attached, determines the magnification. It is usually expressed in millimetres when referring to the eyepiece alone. When interchanging a set of eyepieces on a single instrument, however, some users prefer to refer to identify each eyepiece by the magnification produced.

For a telescope, the angular magnification produced by the combination of a particular eyepiece and objective can be calculated with the following formula:

$$\mathrm{MA} = \frac{f_O}{f_E}$$

where

- MA is the calculated angular magnification.
- f_O is the focal length of the objective.
- f_E is the focal length of the eyepiece, expressed in the same units of measurement as f_T.

Magnification increases, therefore, when the focal length of the eyepiece is shorter or the focal length of the objective is longer. For example, a 25 mm eyepiece in a telescope with a 1200 mm focal length would magnify objects 48 times. A 4 mm eyepiece in the same telescope would magnify 300 times.

Amateur astronomers tend to refer to telescope eyepieces by their focal length in millimetres. These typically range from about 3 mm to 50 mm. Some astronomers, however, prefer to specify the resulting magnification power rather than the focal length. It is often more convenient to express magnification in observation reports, as it gives a more immediate impression of what view the observer actually saw. Due to its dependence on properties of the particular telescope in use, however, magnification power alone is meaningless for describing a telescope eyepiece.

For a compound microscope the corresponding formula is

$$\mathrm{MA} = \frac{DD_{EO}}{f_O f_E} = \frac{D}{f_E} \times \frac{D_{EO}}{f_O}$$

where

- D is the distance of closest distinct vision (usually 250 mm)
- D_{EO} is the distance between the back focal plane of the objective and the back focal plane of the eyepiece (called tube length), typically 160 mm for a modern instrument.
- f_O is the objective focal length and F_E is the eyepiece focal length.

By convention, microscope eyepieces are usually specified by *power* instead of focal length. Microscope eyepiece power P_{E} and objective power P_O are defined by

$$P_E = \frac{D}{f_E}, \quad P_O = \frac{D_{EO}}{f_O}$$

thus from the expression given earlier for the angular magnification of a compound microscope

$$MA = P_E \times P_O$$

The total angular magnification of a microscope image is then simply calculated by multiplying the eyepiece power by the objective power. For example, a 10× eyepiece with a 40× objective will magnify the image 400 times.

This definition of lens power relies upon an arbitrary decision to split the angular magnification of the instrument into separate factors for the eyepiece and the objective. Historically, Abbe described microscope eyepieces differently, in terms of angular magnification of the eyepiece and 'initial magnification' of the objective. While convenient for the optical designer, this turned out to be less convenient from the viewpoint of practical microscopy and was thus subsequently abandoned.

The generally-accepted visual distance of closest focus D is 250 mm, and eyepiece power is normally specified assuming this value. Common eyepiece powers are 8×, 10×, 15×, and 20×. The focal length of the eyepiece (in mm) can thus be determined if required by dividing 250mm by the eyepiece power.

Modern instruments often use objectives optically-corrected for an infinite tube length rather than 160mm, and these require an auxiliary correction lens in the tube.

Location of Focal Plane

In some eyepiece types, such as Ramsden eyepieces (described in more detail below), the eyepiece behaves as a magnifier, and its focal plane is located outside of the eyepiece in front of the field lens. This plane is therefore accessible as a location for a graticule or micrometer crosswires. In the Huygenian eyepiece, the focal plane is located between the eye and field lenses, inside the eyepiece, and is hence not accessible.

Field of View

The field of view, often abbreviated FOV, describes the area of a target (measured as an angle from the location of viewing) that can be seen when looking through an eyepiece. The field of view seen through an eyepiece varies, depending on the magnification achieved when connected to a particular telescope or microscope, and also on properties of the eyepiece itself. Eyepieces are differentiated by their *field stop*, which is the narrowest aperture that light entering the eyepiece must pass through to reach the field lens of the eyepiece.

Due to the effects of these variables, the term "field of view" nearly always refers to one of two meanings:

Actual field of view

the angular size of the amount of sky that can be seen through an eyepiece when used with a particular telescope, producing a specific magnification. It is typically between one tenth of a degree, and two degrees.

Apparent field of view

this is a measure of the angular size of the image viewed through the eyepiece, in other words, how large the image appears (as distinct from the magnification). This is constant for any given eyepiece of fixed focal length, and may be used to calculate what the *actual* field of view will be when the eyepiece is used with a given telescope. The measurement ranges from 35 to over 80 degrees.

It is common for users of an eyepiece to want to calculate the actual field of view, because it indicates how much of the sky will be visible when the eyepiece is used with their telescope. The most convenient method of calculating the actual field of view depends on whether the apparent field of view is known.

If the apparent field of view is known, the actual field of view can be calculated from the following approximate formula:

$$FOV_C = \frac{FOV_P}{mag}$$

or

$$FOV_C = \frac{FOV_P}{\left(\frac{f_T}{f_E}\right)}$$

where

- FOV_C is the actual field of view, calculated in the unit of angular measurement in which FOV_P is provided.
- FOV_P is the apparent field of view.
- *mag* is the magnification.
- f_T is the focal length of the telescope.
- f_E is the focal length of the eyepiece, expressed in the same units of measurement as f_T.

The *focal length* of the telescope objective is the diameter of the objective times the focal ratio. It represents the distance at which the mirror or objective lens will cause light to converge on a single point.

The formula is accurate to 4% or better up to 40° apparent field of view, and has a 10% error for 60°.

If the apparent field of view is unknown, the actual field of view can be approximately found using:

$$FOV_C = \frac{57.3d}{f_T}$$

where

- FOV_C is the actual field of view, calculated in degrees.
- *d* is the diameter of the eyepiece field stop in mm.
- f_T is the focal length of the telescope, in mm.

The second formula is actually more accurate, but field stop size is not usually specified by most manufacturers. The first formula will not be accurate if the field is not flat, or is higher than 60° which is common for most ultra-wide eyepiece design.

Eyepieces for microscopes are usually interchanged to increase or decrease the magnification and to allow the user to select a type with a certain performance characteristic. To allow this eyepieces come in standardized "Barrel diameters".

There are three standard barrel diameters for microscopes. The barrel sizes are usually expressed using inches.

The smallest standard telescope barrel diameter is 0.965 inches (24.5 mm), but has been largely abandoned. The only telescopes still manufactured that use this size are poor-quality telescopes usually found in toy stores and shopping malls. Many of these eyepieces that come with such telescopes are plastic, and some even have plastic lenses. High-quality telescope eyepieces with this barrel size are no longer manufactured:

- The most popular telescope eyepiece barrel diameter is 1¼ inches (31.75 mm). The practical upper limit on focal lengths for eyepieces with 1¼ inch (31.75mm) barrels is about 32 mm. With longer focal lengths, the edges of the barrel itself intrude into the view limiting its size. With focal lengths longer than 32 mm, the available field of view falls below 50°, which most amateurs consider to be the minimum acceptable width. These barrel sizes are threaded to take 30 mm filters (Fig. 3.1).
- Telescope eyepieces with 2 inch (50.8 mm) barrels are also available. The Larger 2 inch (50.8 mm) size helps alleviate the limit on focal lengths. The upper limit of focal length with 2 inch eyepieces is about 50 mm. The trade-off is that these eyepieces are usually more expensive, won't fit in some telescopes, and may be heavy enough to tip the telescope. These barrel sizes are threaded to take 48 mm filters (or rarely 49 mm).

Fig. 3.1. Examples (from left to right) of a 2 inch (50.8 mm), a 1¼ inch (31.75 mm), and a 0.965 inch (24.5mm) eyepiece.

Microscope Eyepieces

Microscopes have standard barrel diameters measured in millimeters: 23.2 mm and 30 mm.

Eye Relief

The eye needs to be held at a certain distance behind the eye lens of an eyepiece to see images properly through it. This distance is called the eye relief. A larger eye relief means that the optimum position is further from the eyepiece, making it easier to view an image. However, if the eye relief is too large it can be uncomfortable to hold the eye in the correct position for an extended period of time, for which reason some eyepieces with long eye relief have cups behind the eye lens to aid the observer in maintaining the correct observing position. The eye pupil should coincide with the Ramsden disc, the image of the entrance pupil, which in the case of an astronomical telescope corresponds to the object glass.

Eye relief typically ranges from about 2 mm to 20 mm, depending on the construction of the eyepiece. Long focal-length eyepieces usually have ample eye relief, but short focal-length eyepieces are more problematic. Until recently, and still quite commonly, eyepieces of a short-focal length have had a short eye relief. Good design guidelines suggest a minimum of 5–6 mm to accommodate the eyelashes of the observer to avoid discomfort. Modern designs with many lens elements, however, can correct for this, and viewing at high power becomes more comfortable. This is especially the case for spectacle wearers, who may need up to 20 mm of eye relief to accommodate their glasses.

Eyepiece Designs

Technology has developed over time and there are a variety of eyepiece *designs* for use with optical telescopes. They vary in their internal lens configuration and different designs are sometimes more appropriate both for different types of viewing, and for different types of telescope. Eyepiece designs include Huygens, Ramsden, Kellner, Orthoscopic, Erfle, König, Plössl, RKE, and Nagler. These are described in more detail below.

Huygens

The two element Huygens eyepiece was invented by Christiaan Huygens in the 17th century. This optical design is now considered obsolete. Their main use in optics is as an example of the simplest possible compound lens design.

Despite being deprecated, these eyepieces are inexpensive to make and so are often sold with the cheapest telescopes and microscopes. Huygens eyepieces suffer from short eye relief, high image distortion (especially on short focus telescopes), chromatic aberration and have very narrow apparent field of view.

Essentially their only good use is for projection of a solar image onto a screen. Because Huygens eyepieces do not contain cement to hold the lens elements, they are less likely to be damaged by the intense, concentrated light of sun. Lens cement can overheat and either dissolve or burn. However, since a Barlow lens can often be used to focus a solar image on a distant screen, and does not form an internal focus to concentrate heat, even this last use is obsolete.

Huygens eyepieces consist of two plano-convex lenses with the plane sides towards the eye separated by an air gap. The lenses are called the eye lens and the field lens. It is usually designed for zero transverse chromatic aberration. The focal plane is located between the two lenses. If the lenses are made of glass of the same refractive index, to be used with a relaxed eye and a telescope with an infinitely distant objective then the separation is given by:

$$d = \frac{1}{2}(f_A + f_B)$$

where f_A and f_B are the focal lengths of the component lenses.

Ramsden

The Ramsden eyepiece, created by astronomical and scientific instrument maker Jesse Ramsden in the 18th century, comprises two plano convex lenses with the same focal length and glass, placed less than one focal length apart. The separation varies between different designs, but is typically somewhere between 7/10 and 7/8 of the focal length of the lenses, the choice being a tradeoff between residual transverse chromatic aberration (at low values) and at high values running the risk of the field lens touching the focal plane when used by an observer who works with a close virtual image such as a myopic observer, or a young person whose accommodation is able to cope with a close virtual image (this is a serious problem when used with a micrometer as it can result in damage to the instrument).

A separation of exactly 1 focal length is also inadvisable since it renders the dust on the field lens disturbingly in focus. The two curved surfaces face inwards. The focal plane is thus located outside of the eyepiece and is hence accessible as a location where a graticule, or micrometer crosshairs may be placed. Because a separation of exactly one focal length would be required to correct transverse chromatic aberration, it is not possible to correct the Ramsden design completely for transverse chromatic aberration. The design is slightly better than Huygens but still not up to today's standards.

It remains highly suitable for use with instruments operating using near monochromatic light sources *e.g.* polarimeters.

Fig. 3.2. A 25 mm Kellner eyepiece.

Kellner or "Achromat"

Carl Kellner designed this first modern achromatic eyepiece in 1850, also called an "achromatized Ramsden". Kellner eyepieces are a 3-lens design. An achromatic doublet is used in place of the eye lens in the Ramsden design to correct the residual transverse chromatic aberration. They are inexpensive and have fairly good image from low to medium power and are far superior to Huygenian or Ramsden design. The biggest problem of Kellner eyepieces was internal reflections. Today's anti-reflection coatings make these usable, economical choices for small to medium aperture telescopes with focal ratio f/6 or longer.

Abbe or "Ortho"

The 4-element Abbe eyepiece was invented by Ernst Abbe in 1880, and is called "*orthoscopic*" or "*orthographic*" because of its low degree of distortion; usually the eyepiece is simply called an "ortho". The Abbe design uses a convex-convex triplet field lens and a convex-flat singlet eye lens. Orthos have nearly perfect image quality and good eye relief, but a little bit narrow apparent field of view — about 40°–45°. Until the advent of multicoatings and the popularity of the Plössl, orthos were the most popular design for telescope eyepieces. Even today these eyepieces are superior to most others for planetary and lunar viewing.

Erfle

Erfles were invented during the first world war for military purposes, described in US patent by Heinrich Erfle number 1,478,704 of Aug 1921. They are a 5-element design which is a logical extension to wider fields of the four lens military eyepiece design. In effect, they are Plössls with extra lenses.

Erfle eyepieces are designed to have wide field of view (about 60 degrees), but they are unusable at high powers because they suffer from astigmatism and ghost images. However, with lens coatings at low powers (focal lengths of 20 mm and up) they are acceptable, and at 40 mm they can be excellent. Erfles are very popular because they have large eye lenses, good eye relief and can be very comfortable to use.

König

The König eyepiece was designed in 1915 by German optician Albert König (1871"1946). The original design is a simplified Abbe, with a leading doublet instead of a triplet. The original design allows for high magnification with remarkably high eye relief — the highest eye relief proportional to focal length of any design before the Nagler, in 1979. The field of view of about 55° makes its performance similar to the Plössl, with the advantage of requiring one less lens.

König's original 1915 form is the simplest, and is composed of two lens groups: a concave-convex positive doublet and a convex~flat positive singlet. The strongly convex surfaces of the doublet and singlet face and (nearly) touch each other. The doublet has its concave surface facing the light source and the singlet has its almost flat (slightly convex) surface facing the eye.

Modern versions of Königs can use improved glass, or add more lenses, grouped into various combinations doublets and singlets. The most typical adaptation is to add a positive, concave-convex simple lens before the doublet, with the concave face towards the light source and the convex surface facing the doublet. Modern improvements typically have fields of view of 60°-70°.

Monocentric

An eyepiece design consisting of crown glass lenses cemented on both sides of a flint glass elements. The elements are thick, strongly curved, and their surfaces have a common centre. The design, like other solid eyepieces such as the Tolles, is free from ghost reflections and gives a bright contrasty image. However, the filed of view is 30°, suitable only for planetary and double-star observations using an tracking telescope.

Plössl

Originally designed by Georg Simon Plössl in 1860, several versions can be found on the amateur astronomy market. By far the Plössl eyepiece is currently the most widely used design. The name Plössl eyepiece covers a range of eyepieces with at least four optical elements. Usually consisting of two sets of doublets (a doublet lens a convex and concave element sandwiched together). The compound Plössl lens provides a large *apparent field of view* along with relatively large FOV. This makes this lens ideal for a variety of observational purposes including deep sky and planetary viewing.

Fig. 3.3. A 25 mm Plössl eyepiece

The chief disadvantage of the Plössl optical design is short eye relief, which is restricted to about 70-80% of focal length. The short eye relief is more critical in short focal lengths, when viewing can become uncomfortable.This eyepiece is one of the more expensive to manufacture because of the quality of glass, and the need for well matched convex and concave lenses to prevent internal reflections. Due to this fact, the quality of different Plössl eyepieces varies. There are notable differences between cheap Plössls with simplest anti-reflection coatings and well made ones.

RKE

An RKE eyepiece is an adaptation of a Kellner eyepiece designed by Dr. David Rank for the Edmund Scientific Corporation, who marketed it throughout the late 1960s and early 1970s. This design provides slightly wider field of view than classic Kellner design.

There is some ambiguity about what *RKE* stands for. According to an e-mail from Edmund, *RKE* stands for *Rank Kellner Eyepiece*. Others speculate it stands for *Rank Kellner Edmund* or *Reversed Kellner Eyepiece*; the latter because the elements within the eyepiece in effect have been reversed from the Kellner design on which it is based. This arrangement makes the design similar to a widely spaced version of the König design.

Nagler

Fig. 3.4. The NAGLER eye piece

Invented by Albert Nagler and patented in 1979, the Nagler eyepiece is a design optimized for astronomical telescopes to give an ultra-wide field of view (82°) that has good correction for astigmatism and other aberrations. This is achieved using exotic high-index glass and up to eight optical elements in 4 or 5 groups; there are 5 similar designs called the *Nagler*, *Nagler type 2*, *Nagler type 4*, *Nagler type 5*, *Nagler type 6*.

The number of elements in a Nagler makes them seem complex, but the idea of the design is fairly simple: every Nagler has a negative doublet field lens, which increases magnification, followed by several positive groups. The positive groups, considered separate from the first negative group, combine to have long focal length, and form a positive lens. That allows the design to take advantage of the many good qualities of low power lenses. In effect, a Nagler is a superior version of a Barlow lens combined with a long focal length eyepiece. This design has been widely copied in other wide field or long eye relief eyepieces.

The main disadvantage to Naglers is in their weight. Long focal length versions exceed 0.5 kg, which is enough to unbalance many microscopes. Amateurs fondly refer to Naglers as "paperweights", because of their heft, or "hand grenades", because of their size and shape. Another disadvantage is a high purchase cost, with large Naglers' prices comparable to the cost of a small microscope. Hence these eyepieces are regarded by many amateur microbiologists as a luxury.

Objectives

The central piece of an optical microscope is its objective lens. Most objective lenses should strictly be called 'objective lens arrays', since they are made out of many lenses - but no-one calls them that.

The lenses that make up the objective lens array are commonly encased in a metal cylinder a few centimetres in height, and about one centimetre in diameter. Such a cylinder is often termed an 'objective barrel'. At the top and bottom of the barrel are the lenses (and there can be many more lenses arranged in complicated ways in the inner part of an objective barrel). At the top of the barrel there is also a screw with which one can attach the objective lens to the microscope (normally on a revolving objective drum). The bottom end of the barrel will be nearest to the object to be inspected. Specifications are engraved on the side of the cylinder, giving information about the type, the magnification power and the resolution of the objective lens.An objective lens is very generally defined as the first lens of any optical apparatus from the perspective of the object (hence the name). The term is not confined to microscopy, and can be found in other fields of optical science; this Guide Entry, however, describes the aspects of objective lenses concerning microscopy in particular.

The Objective Lens

In microscopy there are different types of objective lens. Their general appearance is more or less the same; but their magnification can vary, and some special objectives require extra handling and adjustment, while some are more simple. Above is an illustration of a generic microscope objective lens array. In the example shown, the objective lens is a compound of two lenses which work together as if they were just one stronger lens. Many of the microscope objective lenses are compounded of many weaker lenses to form one strong lens.

Magnification, Numeric Aperture and Resolution

Strangely, the magnification factor is the least important property of an objective lens. It's merely equivalent to the strength of the lens array. The bigger the magnification factor, the more the user will be able to zoom in. But the ability to zoom in is of little help if the resolution is low. On the other hand, high resolution makes little sense unless one is able to zoom in.The resolution is defined by the smallest distance between two dots that still allows them to be recognised as two distinct points. The problem with microscopy is that these dots have a minimum size, and this size depends in its turn on the sharpness of the dots. So the crucial question is: How sharp can images get? Or, the other way round: How blurry do dots get?The reason for the blurriness lies in the fact that not all of the light from any one point of the sample can be collected. The objective lens has a finite width, and it is placed at a finite distance from the observed dot. Hence only a cone of light can be collected by the objective lens. And it's not the loss of intensity but the loss of directional information that is the trouble. The problem lies in the missing interference pattern at the edge of the projected dot - it gets blurred.

As if that was not enough, the missing interference also depends on the colour of the light. So in the end, the blurriness and thus the resolution depends on the wavelength of the light used, and the width of the collection cone. The distance between the two dots decreases with the cone size (Numeric Aperture) and increases with the wavelength. The rough formula is: wavelength of the light (in hundreds of nanometers) divided by roughly twice the numeric aperture (which is around unity). So - even more roughly - the resolution is about half the wavelength.

Increasing the Numeric Aperture

There is a neat little effect in optics called Total Internal Reflection. This is what makes the surface of a lake or a glass plate look like a mirror at shallow angles. The angle where light starts to get reflected (the Critical Angle) is proportional to the quotient of the refractive indices of the media in question. If this quotient is high (eg going from water to air) then the angles get less shallow, if the quotient is low (eg going from glycerine to glass) then the angles get wider. Now how does this affect numeric aperture? How can the detection cone get wider? The answer is: When the refractive indices of the medium between the object and the first lens come close. On the other hand, because of this, the magnification power of this lens will diminish. There is an optimal compromise between losing lens power and increasing the numeric aperture. For that reason some objectives work in special immersion media, like special immersion oils or water. These objective lenses have a very complex design to optimise the effect of the immersion medium.

Types of Objective

There are two main types of objective: **'immersion' objectives** and **'air' objectives**. These can both be achromatic and/or aspheric. One further exotic type is the 'mirror' objective, which works in the same way as a mirror telescope, but is a lot more difficult to handle. It has the advantage that it doesn't lose as much light-intensity as 'lens' objectives. All of these objectives are available in further, more specialized variations, which allow a finer tuning of optical possibilities.

In this chapter we will explore the technical aspects of the heart of the optical microscope- the *objective lens*. The evolution of the objective in the past decade has been exceptional. Driven by thedemands of biological, biomedical, and material science researchers coupled with advances in lenscomputer design, lens coating technology, and glass formulation chemistry, this evolution continuesFirst we will briefly review some historical background, followed by current terminology, and endingwith objective oriented optical techniques.Early nineteenth century microscopists were plagued by a lack of image quality including blurringand various chromatic, spherical and axis aberrations caused by design flaws and glass raw materialquality. In the mid nineteenth century work by Lister and Amici reduced chromatic aberration andraised the numerical aperture by introducing the first achromatic objectives. In the last part of thenineteenth century Ernst Abbe and Carl Zeiss collaborated to produce the first apochromaticobjectives. These apochromats again increased numerical aperture, reduced spherical aberration, andmost importantly corrected for chromatic aberration almost perfectly.In order to fully understandobjectives some basic terms need to be discussed briefly.

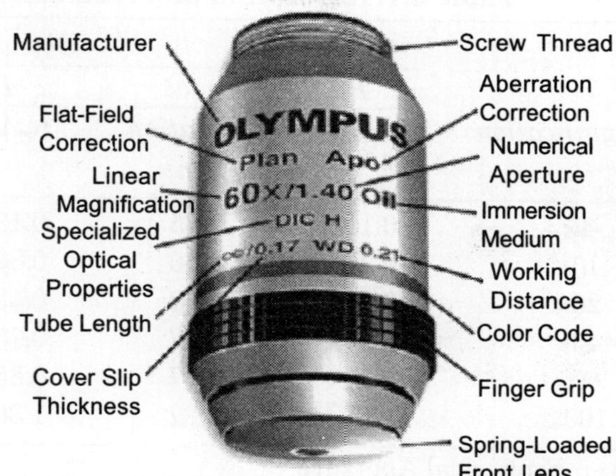

Fig . 3.5. The Specification of objectives

Numericalaperture (NA) simply defined, is the ability of an objective to resolve fine detail coupled with itsability to gather light expressed as a number. The higher the NA of an objective the greater theresolution and brightness of the observed image. All dry objectives have NA's lower than 1 whereas oil objectives generally greater than 1. Objectives are made of various types of glass. The simplesttype of objective, Achromat, is made of crown and flint glass elements. The next level of objective iscalled a Fluorite is composed of elements of calcium fluorite and is some cases quartz glasspermitting transmission in the UV range (340nm). The most highly corrected objective is the Apochro-mat. It is composed of elements of fluorite and apochromatic glass. Still other objectives arecapable of wavelength trans-mission of 400-1200 nm by using proprietary coating, glass elements, andcementing agents.The use of various devices that interfere with the light path constructively by enhancing contrast hascontinued to increase in usage in the last several decades. Specially modified objectives like phase,Nomarski, Hoffman and polarized light are necessary for these techniques to even happen. Light-sources (sometimes used with

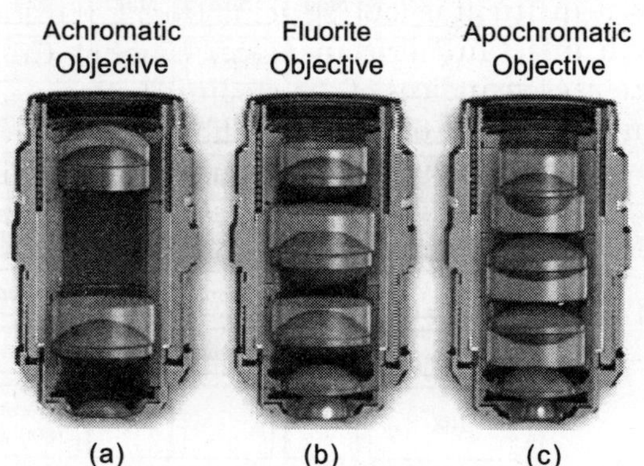

Fig. 3.6. The Optical correction in objectives (*a*) Achromatic objectives, the lowest level of correction, contain two doublets and a single front lens; (*b*) Fluorites orsemiapochromatic objectives, a medium level of correction, contain three doublets, a meniscus lens, and a single front lens; and (*c*) Apochromatic objectives, the highest level of correction, contain a triplet, two doublets, a meniscus lens, and a singlehemispherical front lens.

interference filters and dichroic mirrors) for image enhancement suchas mercury, xenon, IR, tin halide, and lasers are also expanding the use of specialty objectives.

Table 3.1. Resolution and Numerical Aperture by Objective Type

	Objective Type					
	Plan Achromat		*Plan Fluorite*		*Plan Apochromat*	
Magnification	*N.A.*	*Resolution (μm)*	*N.A.*	*Resolution (μm)*	*N.A.*	*Resolution (μm)*
4×	0.10	2.75	0.13	2.12	0.20	1.375
10×	0.25	1.10	0.30	0.92	0.45	0.61
20×	0.40	0.69	0.50	0.55	0.75	0.37
40×	0.65	0.42	0.75	0.37	0.95	0.29
60×	0.75	0.37	0.85	0.32	0.95	0.29
100×	1.25	0.22	1.30	0.21	1.40	0.20

N.A. = Numerical Aperture

Table 3.2.

Objective Lenses	Eyepiece Lenses							
	5X		10X		15X		20X	
	Eyepieces		Eyepieces		Eyepieces		Eyepieces	
	Total Mag.	Field Size	Total Mag.	Field Size	Total Mag.	Field Size	Total Mag.	Field Size
1X	5X	22mm	10X	20mm	15X	13mm	20X	9.5mm
2X	10X	11mm	20X	10mm	30X	6.5mm	40X	4.8mm
1X	5X	22mm	10X	20mm	15X	13mm	20X	9.5mm
3X	15X	7.3mm	30X	6.7mm	45X	4.3mm	60X	3.2mm
2X	10X	11mm	20X	10mm	30X	6.5mm	40X	4.8mm
4X	20X	5.5mm	40X	5mm	60X	3.2mm	80X	2.4mm
10X			100X	1.8mm 1800um				
40X			400X	0.45mm 450um				
100X			1000X	0.18mm 180um				

High power microscopes usually have 10X eyepiece lenses so we only list one column above the 4X objective. If you wanted to look at an amoeba that is 150um long (0.15mm) then you would use either the 40X or 100X objectives.. There are 25 mm in an inch and there are 1,000 um (micrometers) in a single millimeter.

Oil Immersion Lens and its Uses

In light microscopy, **oil immersion** is a technique used to increase the resolution of a microscope. This is achieved by immersing both the objective lens and the specimen in a transparent oil of high refractive index, thereby increasing the numerical aperture of the objective lens.

Immersion oils are transparent oils that have specific optical and viscosity characteristics necessary for use in microscopy. An **oil immersion objective** is an objective lens specially designed to be used in this way. Many condensers also give optimal resolution when the condenser lens is immersed in oil.

The resolution of a microscope is defined as the minimum separation needed between two objects under examination in order for the microscope to discern them as separate objects. This minimum distance is labeled δ. If two objects are separated by a distance shorter than δ, they will appear as a single object in the microscope.

A measure of the resolving power of a lens is given by its numerical aperture, NA:

$$\delta = \frac{\lambda}{NA}$$

where λ is the wavelength of light. From this it is clear that a good resolution (small ä) is connected with a high numerical aperture.

The numerical aperture of a lens is defined as

$$NA = nsin\alpha_0$$

where α_0 is the angle spanned by the objective lens seen from the sample, and n is the refractive index of the medium between the lens and specimen (≈1 for air). State of the art objectives can have a numerical aperture of up to 0.95. Because sin α_0 is always less than or equal to unity, the numerical aperture can never be greater than unity for an objective lens in air. If the space between the objective lens and the specimen is filled with oil however, the numerical aperture can obtain values greater than unity. This is because oil has a refractive index greater than 1.

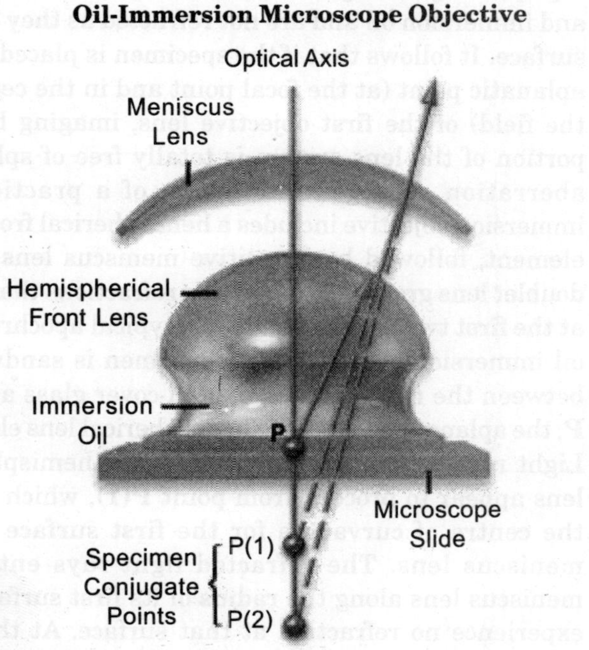

Oil-Immersion Microscope Objective

Fig. 3.7. The Mechanism of Oil Immersion Lens.

Oil immersion Objectives

From the above it is understood that oil between the specimen and the objective lens improves the resolving power by a factor $1/n$. Objectives specifically designed for this purpose are known as oil immersion objectives. Oil immersion objectives are used only at very large magnifications that require high resolving power. Objectives with high power magnification have short focal lengths, facilitating the use of oil. The oil is applied to the specimen (conventional microscope), and the stage is raised, immersing the objective in oil. (In inverted microscopes the oil is applied to the objective).The refractive indices of the oil and of the glass in the first lens element are nearly the same, which means that the refraction of light will be small upon entering the lens (the oil and glass are optically very similar). Oil immersion objectives are designed with this in mind, and can not be used without oil, because in this case there will be much refraction of light at the glass/air interface, drastically changing the path of light entering the lens and making it impossible to achieve focus. Nevertheless, in situations in which not maximal clarity but a closer general view is preferred (for instance, where magnifications must be changed back and forth during the observation of a delicately mounted slide), the oil can be skipped at the cost of image sharpness. Using the oil immersion objective with oil can only be done on fixed specimens.Another advantage of using oil is that it reduces reflective losses as light enters the lens (again because the oil and glass are optically alike).

Objective numerical aperture can be dramatically increased by designing the objective to be used with an immersion medium, such as oil, glycerin, or water. By using an immersion medium with a refractive index similar to that of the glass coverslip, image degradation due to thickness variations of the cover glass are practically eliminated whereby rays of wide obliquity no longer undergo refraction and are more readily grasped by the objective. Typical immersion oils have a refractive index of 1.51 and a dispersion similar to that of glass coverslips. Light rays passing through the specimen encounter a homogeneous medium between the coverslip and immersion oil and are not refracted as they enter the lens, but only as they leave its upper surface. It follows that if the specimen is placed at the aplanatic point (at the focal point and in the centre of the field) of the first objective lens, imaging by this portion of the lens system is totally free of spherical aberration. The general design of a practical oil immersion objective includes a hemispherical front lens element, followed by a positive meniscus lens and a doublet lens group. The aplanatic refractions that occur at the first two lens elements in a typical apochromatic oil immersion objective. The specimen is sandwiched between the microscope slide and cover glass at point **P**, the aplanatic point of the hemispherical lens element. Light rays refracted at the rear of the hemispherical lens appear to proceed from point **P(1)**, which is also the centre of curvature for the first surface of the meniscus lens. The refracted light rays enter the meniscus lens along the radius of its first surface and experience no refraction at that surface. At the rear surface of the meniscus lens, light rays are refracted aplanatically, so they appear to diverge from point **P(2)**.

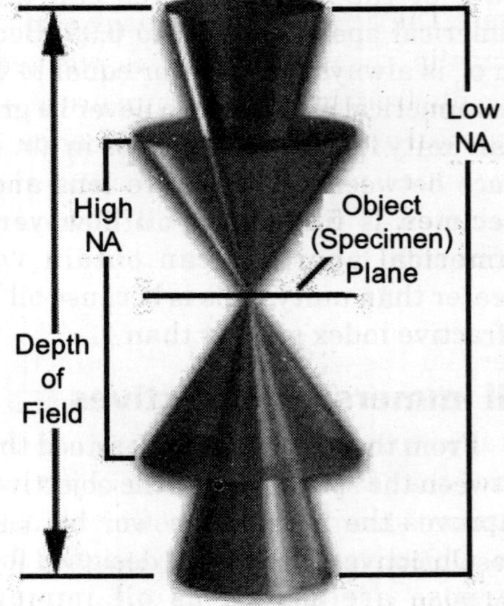

Fig. 3.8.

Refraction of the light rays at the surfaces of subsequent lens groups in the objective complete the convergence of light rays originating from point **P**, thus forming the intermediate image.

Properly designed oil immersion objective lenses also correct for chromatic defects that are introduced by the first two lens elements, while introducing a minimum amount of spherical aberration. The fact that the light cone is partially converged before entering the first lens element aids in the control of spherical aberration. It should be noted that employing an oil immersion objective without the application oil between the coverslip and first lens element results in defective images. This due to refraction that occurs at the surface of the front lens, which introduces spherical aberration that cannot be corrected by subsequent lens components within the objective.

The advantages of oil immersion objectives are severely compromised if the wrong immersion fluid is utilized. Microscope manufacturers produce objectives with tight tolerances to refractive index and dispersion, which require matching values in the liquid placed between the cover glass and objective front lens. It is advisable to employ only the oil intended by the objective manufacturer, and to not mix immersion oils between manufacturers to avoid unpleasant artifacts such as crystallization or phase separation.

Objectives that use water and/or glycerin as an imaging medium are also available for applications with living cells in culture or sections of tissue immersed in physiological saline solution. Plan apochromat water immersion lenses are equipped with correction collars and numerical apertures up to 1.2, slightly less than their oil immersion counterparts. These objectives allow microscopists to focus through up to 200 microns of aqueous media and still retain excellent optical correction. The downside is that high numerical aperture water immersion lenses often cost many thousands of dollars and the image can still degrade when the objective is focused deeply through refractile tissue or cell parts. For more details on water, glycerin, and oil immersion objectives, visit our section on immersion media in the microscopy primer.The axial range, through which an objective can be focused without any appreciable change in image sharpness, is referred to as the objective **depth of field**. This value varies radically from low to high numerical aperture objectives, usually decreasing with increasing numerical aperture. At high numerical apertures, the depth of field is determined primarily by wave optics, while at lower numerical apertures, the geometrical optical "circle of confusion" dominates. The total depth of field is given by the sum of the wave and geometrical optical depths of field as:

$$\mathbf{d_{tot} = ln/NA^2 + (n/M \bullet NA)e}$$

where **l** is the wavelength of illumination, **n** is the refractive index of the imaging medium, **NA** is the objective numerical aperture, **M** is the objective lateral magnification, and **e** is the smallest distance that can be resolved by a detector that is placed in the image plane of the objective. Notice that the diffraction-limited depth of field (the first term on the right-hand side of the equation) shrinks inversely with the square of the numerical aperture, while the lateral limit of resolution is reduced with the first power of the numerical aperture. The result is that axial resolution and the thickness of optical sections are affected by the system numerical aperture much more than is the lateral resolution of the microscope (see Table 3.3).

Table 3.3. Depth of Field and Image Depth Table

Magnification	*Numerical Aperture (mm)*	*Depth of Image*	
		Field (mm)	*Depth*
4×	0.10	15.5	0.13
10×	0.25	8.5	0.80
20×	0.40	5.8	3.8
40×	0.65	1.0	12.8
60×	0.85	0.40	29.8
100×	0.95	0.19	80.0

Use of Oil Immersion (100×)

Materials

- Microscope equipped with 100X, oil immersion lens.
- Immersion oil.
- Prepared slide of bacteria or suitably small objects.

Procedure

1. Place the prepared slide on the microscope stage. Using the 10X objective, focus the microscope on an appropriate field containing bacteria. Centre the bacteria in the field of view, by manipulating the lateral movement knobs.
2. Rotate the nosepiece to the 40X objective and refocus with the fine focus. Again, centre the object which you wish to examine.
3. Rotate the nosepiece so that an intermediate position between the 40X and 100X objectives is obtained.
4. Place a small drop of immersion oil 10 on the centre of the viewing area of the slide.
5. Continue to rotate the nosepiece so that the 100X objective is rotated into the oil.

   Do not under any circumstances place the 40x objective in the oil

   Use only the fine focus and refocus your specimen.
6. Determine the shapes of the bacteria and draw them in the place provided.
7. Immediately after using the oil, remove any residual oil from the slide and from the front of the 100X objective by gently rubbing with lens paper dipped in xylol, toluene, or better, a xylol substitute designed for this purpose.

Nearly all organic solvents, and especially xylol and toluene, are potentially hazardous. They are flammable and readily enter the boddy by direct absorption through the skin or lungs. Consistent high exposure to thses solvents has been linked with liver damage and potential carcinogenesis. Use all solvents sparingly when needed and always in a well-ventilated area.

Once oil is placed on the slide, the 10X and 40X objectives will no longer be useful until the oil is removed. The oil will blur any image attempted with the lower magnifications. Use oil only when necessary and after completing all work at the lower magnifications. In order to return to work at the lower magnifications, the slide must be completely cleaned of any residual oil and dried. For this reason, oil immersion is employed only when necessary, and only after thorough observations at the high dry magnification (40X). 11

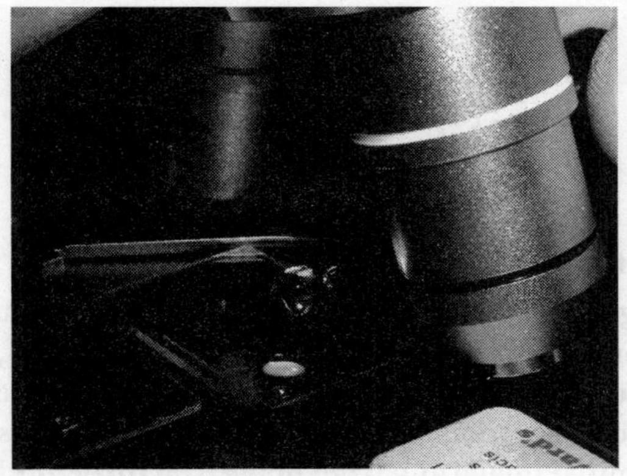

Fig. 3.9. The Use of Immersion Oil Immersion lens

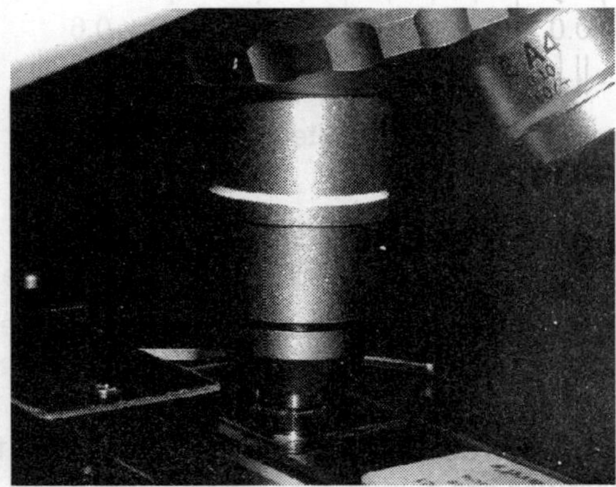

Fig. 3.10. Working Oil Immersion Lens

Identification of the properties of individual objectives is usually very easy because important parameters are often inscribed on the outer housing (or barrel) of the objective itself. Typical 60x plan apochromat objective, including common engravings that contain all of the specifications necessary to determine what the objective is designed for and the conditions necessary for proper use.

Ocular and Stage Micrometers for Measurements of the Specimens

To measure an object seen in a microscope, an ocular micrometer serves as a scale or rule. This is simply a disc of glass upon which equally spaced divisions are etched. The rule may be divided into 50 subdivisions, or more rarely 100 subdivisions. To use the ocular micrometer, calibrate it against a fixed and known ruler, the stage micrometer. Stage micrometers also come in varying lengths, but most are 2 mm long and subdivided into 0.01 mm (10 micrometer) lengths. Each objective will need to be calibrated independently. To use, simply superimpose the ocular micrometer onto the stage micrometer and note the relationship of the length of the ocular to the stage micrometer (Fig. 3.11). Note that at different magnifications, the stage micrometer changes, but the ocular micrometer is fixed in dimension. In reality, the stage micrometer is also fixed, and what is changing is the power of the magnification of the objective.

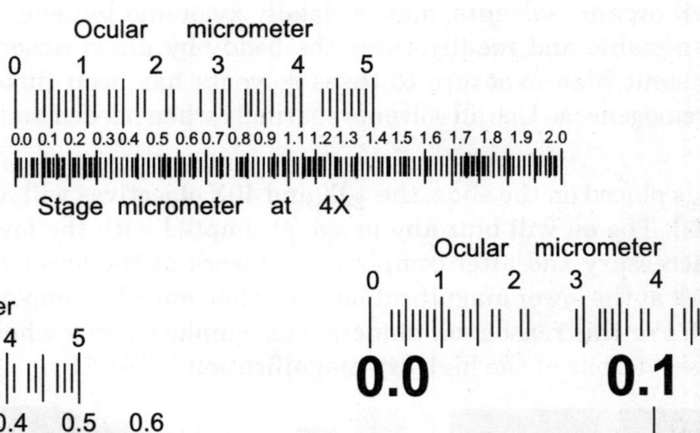

Fig. 3.11. The Stage and the Ocular Micrometer

Microscope manufacturers offer a wide range of objective designs to meet the performance needs of specialized imaging methods, to compensate for cover glass thickness variations, and to increase the effective working distance of the objective. Often, the function of a particular objective is not obvious simply by looking at the construction of the objective. Finite microscope objectives are designed to project a diffraction-limited image at a fixed plane (the **intermediate** image plane), which is dictated by the microscope tube length and located at a pre-specified distance from the rear focal plane of the objective. Microscope objectives are usually designed to be used with a specific group of oculars and/or tube lenses strategically placed to assist in the removal of residual optical errors. As an example, older Nikon and Olympus compensating eyepieces were used with high numerical aperture fluorite and apochromatic objectives to eliminate lateral chromatic aberration and improve flatness of field. Newer microscopes (from Nikon and Olympus) have objectives that are fully corrected and do not require additional corrections from the eyepieces or tube lenses.

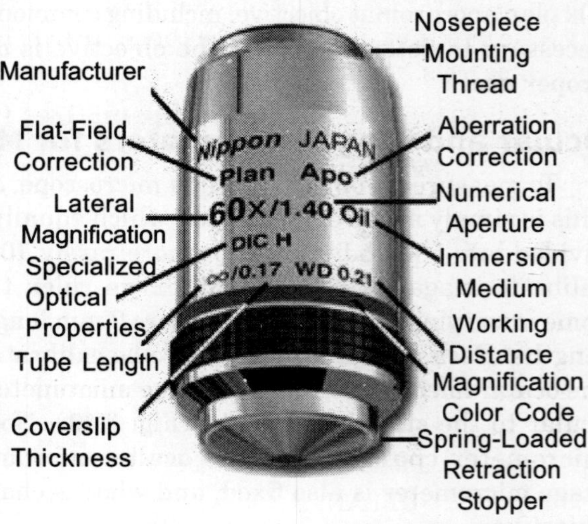

Fig. 3.12. 60× Plan Apochromat Objective.

Most manufacturers have now transitioned to infinity-corrected objectives that project emerging rays in parallel bundles from every azimuth to infinity. These objectives require a tube lens in the light path to bring the image into focus at the intermediate image plane. Infinity-corrected and finite-tube length microscope objectives are not interchangeable and must be matched not only to a specific type of microscope, but often to a particular microscope from a single manufacturer. For example, Nikon infinity-corrected objectives are ***not*** interchangeable with Olympus infinity-corrected objectives, not only because of tube length differences, but also because the mounting threads are not the same pitch or diameter. Objectives usually contain an inscription denoting the tube focal length as will be discussed.

There is a wealth of information inscribed on the barrel of each objective, which can be broken down into several categories. These include the linear magnification, numerical aperture value, optical corrections, microscope body tube length, the type of medium the objective is designed for, and other critical factors in deciding if the objective will perform as needed. A more detailed discussion of these properties is provided below and in links to other pages dealing with specific issues.

- **Manufacturer** - The name of the objective manufacturer is almost always included on the objective. The objective illustrated in Figure 1 was made by a fictitious company named **Nippon** from Japan, but comparable objectives are manufactured by **Nikon**, **Olympus**, **Zeiss**, and **Leica**, companies who are some of the most respected manufacturers in the microscope business.
- **Linear Magnification** - In the case of the apochromatic objective in Figure 1, the linear magnification is 60x, although the manufacturers produce objectives ranging in linear magnification from 0.5x to 250x with many sizes in between.
- **Optical Corrections** - These are usually listed as **Achro** and **Achromat** (achromatic), as **Fl**, **Fluar**, **Fluor**, **Neofluar**, or **Fluotar** (fluorite) for better spherical and chromatic corrections, and as **Apo** (apochromatic) for the highest degree of correction for **spherical and chromatic aberrations. Field curvature** corrections are abbreviated **Plan**, **Pl**, **EF**, **Achroplan**, **Plan Apo**, or **Plano**. Other common abbreviations are **ICS** (infinity corrected system) and **UIS** (universal infinity system), **N** and **NPL** (normal field of view plan), **Ultrafluar** (fluorite objective with glass that is transparent down to 250 nanometers), and **CF** and **CFI** (chrome-free; chrome-free infinity). The objective in the illustration (Figure 1) is a plan apochromat that enjoys the highest degree of optical correction. Table 3.4 for a complete list of abbreviations often found inscribed on objective barrels.

Table 3.4. Specialized Objective Designations

Abbreviation	*Type*
Achro, Achromat	Achromatic aberration correction
Fluor, Fl, Fluar, Neofluar, Fluotar	Fluorite aberration correction
Apo	Apochromatic aberration correction
Plan, Pl, Achroplan, Plano	Flat Field optical correction
EF, Acroplan	Extended Field(field of view less than Plan)
N, NPL	Normal field of view plan
Plan Apo	Apochromatic and Flat Field correction
UPLAN	Olympus Universal Plan (Brightfield, Darkfield, DIC, and Polarized Light)

Abbreviation	*Type*
LU	Nikon Luminous Universal (Brightfield, Darkfield, DIC, and Polarized Light)
L, LL, LD, LWD	Long Working Distance
ELWD	Extra-Long Working Distance
SLWD	Super-Long Working Distance
ULWD	Ultra-Long Working Distance
Corr, W/Corr, CR	Correction Collar
I, Iris, W/Iris	Adjustable numerical aperture(with iris diaphragm)
Oil, Oel	Oil Immersion
Water, WI, Wasser	Water Immersion
HI	Homogeneous Immersion
Gly	Glycerin Immersion
DIC, NIC	Differential orNomarski Interference Contrast
CF, CFI	Chrome-Free,Chrome-Free Infinity-Corrected (Nikon)
ICS	Infinity Color-Corrected System (Zeiss)
RMS	Royal Microscopical Societyobjective thread size
M25, M32	Metric 25-mm objective thread;Metric 32-mm objective thread
Phase, PHACO, PC	Phase Contrast
Ph 1, 2, 3, etc.	Phase Condenser Annulus 1, 2, 3, etc.
DL, DLL, DM, BM	Phase Contrast: Dark Low, Dark Low Low, Dark medium, Bright Medium
PL, PLL	Phase Contrast: Positive Low, Positive Low Low
PM, PH	Phase Contrast: Positive Medium, Positive High Contrast (Regions with higherrefractive index appear darker.)
NL, NM, NH	Phase Contrast: Negative Low, Negative Medium, Negative High Contrast(Regions with higherrefractive index appear lighter.)
P, Po, Pol, SF	Strain-Free, Low Birefringence,for Polarized Light
U, UV, Universal	UV transmitting(down to approximately 340 nm) for UV-excited epifluorescence
M	Metallographic (no coverslip)
NC, NCG	No Coverslip
EPI	Oblique or Epi illumination
TL	Transmitted Light
BBD, HD, B/D	Bright or Dark Field (Hell, Dunkel)
D	Darkfield
H	For use with a heating stage
U, UT	For use with a universal stage
DI, MI, TI	Interferometry, Noncontact,Multiple Beam (Tolanski)

- **Numerical Aperture** - This is a critical value that indicates the light acceptance angle, which in turn determines the light gathering power, the resolving power, and depth of field of the objective.

  Some objectives specifically designed for transmitted light fluorescence and darkfield imaging are equipped with an internal iris diaphragm that allows for adjustment of the effective numerical aperture. Abbreviations inscribed on the barrel for these objectives include **I**, **Iris**, and **W/Iris**. The 60x apochromat objective illustrated above has a numerical aperture of 1.4, one of the highest attainable in modern microscopes using immersion oil as an imaging medium.

- **Mechanical Tube Length** - This is the length of the microscope body tube between the nosepiece opening, where the objective is mounted, and the top edge of the observation tubes where the oculars (eyepieces) are inserted. This aspect of microscope design is discussed in more thoroughly in our **mechanical tube length** section of the primer. Tube length is usually inscribed on the objective as the size in number of millimeters (160, 170, 210, etc.) for fixed lengths, or the infinity symbol (¥) for infinity-corrected tube lengths. The objective illustrated in Figure 1 is corrected for a tube length of infinity, although many older objectives will be corrected for tube lengths of either 160 (Nikon, Olympus, Zeiss) or 170 (Leica) millimeters.

- **Cover Glass Thickness** - Most transmitted light objectives are designed to image specimens that are covered by a cover glass (or cover **slip**). The thickness of these small glass plates is now standardized at 0.17 mm for most applications, although there is often some variation in thickness within a batch of coverslips. For this reason, some of the more advanced objectives have a **correction collar** adjustment of the internal lens elements to compensate for this variation. Abbreviations for the correction collar adjustment include **Corr**, **w/Corr**, and **CR**, although the presence of a movable, knurled collar and graduated scale is also an indicator of this feature.

  The interactive Java tutorial linked above allows the visitor to adjust the correction collar on a microscope objective. There are some applications that do not require objectives to be corrected for cover glass thickness. These include objectives designed for reflected light metallurgical specimens, tissue culture, integrated circuit inspection, and many other applications that require observation with no compensation for a cover glass.

- **Working Distance** - This is the distance between the objective front lens and the top of the cover glass when the specimen is in focus. In most instances, the working distance of an objective decreases as magnification increases. Working distance values are not included on all objectives and their presence varies depending upon the manufacturer. Common abbreviations are: **L**, **LL**, **LD**, and **LWD** (long working distance), **ELWD** (extra-long working distance), **SLWD** (super-long working distance), and **ULWD** (ultra-long working distance). Newer objectives often contain the size of working distance (in millimeters) inscribed on the barrel. The objective illustrated in Figure 1 has a very short working distance of 0.21 millimeters.

- **Specialized Optical Properties** - Microscope objectives often have design parameters that optimize performance under certain conditions. For example, there are special objectives designed for polarized illumination signified by the abbreviations **P**, **Po**, **POL**, or **SF** (strain-free and/or having all barrel engravings painted red), phase contrast (**PH**, and/or green barrel engravings), differential interference contrast (**DIC**), and many other abbreviations for additional applications. A list of several abbreviations, often manufacturer specific, is presented in Table 3.4. The apochromat objective illustrated in Table 3.5 is optimized for DIC photomicrography and this is indicated on the barrel. The capital **H** beside the DIC marking indicates that the objective must be used with a specific DIC Wollaston prism optimized for high-magnification applications.

Table 3.5. Objective Numerical Aperture and Working Distance

Optical Correction and Magnification*	*Numerical Aperture*	*Working Distance (Millimeters)*
ACH 10x	0.25	6.10
ACH 20x	0.40	3.00
ACH 40x	0.65	0.45
ACH 60x	0.80	0.23
ACH 100x (Oil)	1.25	0.13
PL 4x	0.10	22.0
PL 10x	0.25	10.5
PL 20x	0.40	1.20
PL 40x	0.65	0.56
PL 100x (Oil)	1.25	0.15
PL FL 4x	0.13	17.0
PL FL 10x	0.30	10.00
PL FL 20x	0.50	1.60
PL FL 40x	0.75	0.51
PL FL 100x (Oil)	1.30	0.10
PL APO 1.25x	0.04	5.1
PL APO 2x	0.06	6.20
PL APO 4x	0.16	13.00
PL APO 10x	0.40	3.10
PL APO 20x	0.70	0.65
PL APO 40x	0.85	0.20
PL APO 60x (Oil)	1.40	1.10
PL APO 100x (Oil)	1.40	0.10

*Abbreviations:
ACH, Achromat
PL, Plan Achromat
PL FL, Plan Fluorite
PL APO, Plan Apochromat

- **Objective Screw Threads** - The mounting threads on almost all objectives are sized to standards of the Royal Microscopical Society (RMS) for universal compatibility. The objective in Figure 1 has mounting threads that are 20.32 mm in diameter with a pitch of 0.706, conforming to the RMS standard. This standard is currently used in the production of infinity-corrected objectives by manufacturers Olympus and Zeiss. Nikon and Leica have broken from the standard with the introduction of new infinity-corrected objectives that have a wider mounting thread size, making Leica and Nikon objectives usable only on their own microscopes. Nikon's reasoning is explained in our section describing the **Nikon CFI60 200/60/25 Specification** for biomedical microscopes. Abbreviations commonly used to denote thread size are: **RMS** (Royal Microscopical Society objective thread), **M25** (metric 25-millimeter objective thread), and **M32** (metric 32-millimeter objective thread).

- **Immersion Medium** - Most objectives are designed to image specimens with air as the medium between the objective and the cover glass.

To attain higher working numerical apertures, many objectives are designed to image the specimen through another medium that reduces refractive index differences between glass and the imaging medium. High-resolution plan apochromat objectives can achieve numerical apertures up to 1.40 when the immersion medium is special oil with a refractive index of 1.51. Other common immersion media are water and glycerin. Objectives designed for special immersion media usually have a color-coded ring inscribed around the circumference of the objective barrel as listed in Table 3 and described below. Common abbreviations are: **Oil, Oel** (oil immersion), **HI** (homogeneous immersion), **W, Water, Wasser** (water immersion), and **Gly** (glycerol immersion).

- **Color Codes** - Microscope manufacturers label their objectives with color codes to help in rapid identification of the magnification and any specialized immersion media requirements. The dark blue color code on the objective illustrated in Figure 1 indicates the linear magnification is 60x. This is very helpful when you have a nosepiece turret containing 5 or 6 objectives and you must quickly select a specific magnification. Some specialized objectives have an additional color code that indicates the type of immersion medium necessary to achieve the optimum numerical aperture. Immersion lenses intended for use with oil have a black color ring, and those intended for use with glycerin have an orange ring. Objectives designed to image living organisms in aqueous media are designated ***water immersion*** objectives with a white ring, and highly specialized objectives for unusual immersion media are often engraved with a red ring. Table 3.6 lists current magnification and imaging media color codes in use by most manufacturers.

Table 3.6. Objective Color Codes

Magnification	*Color Code*
1/2x	No Color Assigned
1x	Black
1.25x	Black
1.5x	Black
2x	Brown (or Orange)
2.5x	Brown (or Orange)
4x	Red
5x	Red
10x	Yellow

Magnification	*Color Code*
16x	Green
20x	Green
25x	Turquoise
32x	Turquoise
40x	Light Blue
50x	Light Blue
60x	Cobalt Blue
63x	Cobalt Blue
100x	White
150x	White
250x	White
Immersion Media	Color Code
Oil	Black
Glycerol	Orange
Water	White
Special	Red

Special Features - Objectives often have additional special features that are specific to a particular manufacturer and type of objective. The plan apochromat objective has a spring-loaded front lens to prevent damage when the objective is accidentally driven onto the surface of a microscope slide.

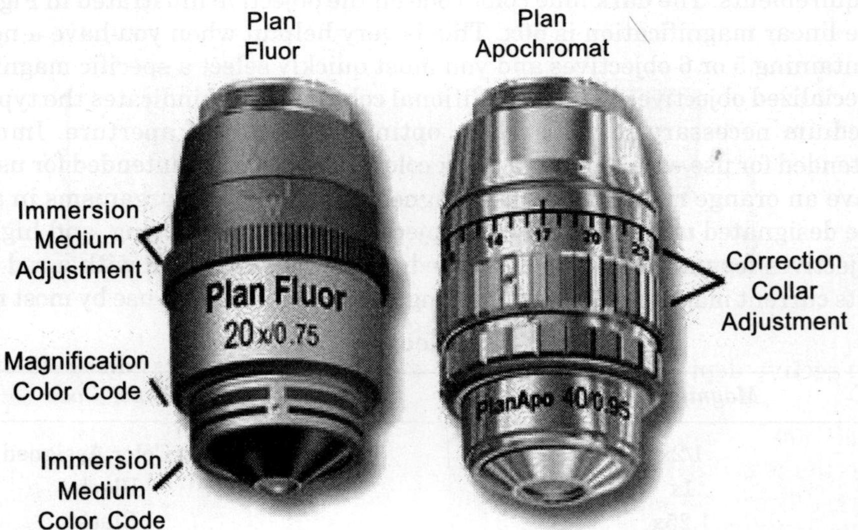

Fig. 3.13.

Other features found on specialized objectives are variable working distance (**LWD**) and numerical aperture settings that are adjustable by turning the correction collar on the body of the objective as illustrated in Fig. 13. The plan fluor objective on the left has a variable immersion medium/numerical aperture setting that allows the objective to be used with both air and an alternative liquid immersion medium, glycerin. The plan apo objective on the right has an adjustable working distance control (termed a "correction collar") that allows the objective to image specimens through glass coverslips of variable thickness. This is especially important in dry objectives with high numerical aperture that are particularly susceptible to spherical and other aberrations that can impair resolution and contrast when used with a cover glass whose thickness differs from the specified design value.

Although not common today, other types of adjustable objectives have been manufactured in the past. Perhaps the most interesting example is the compound "zoom" objective that has a variable magnification, usually from about 4x to 15x. These objectives have a short barrel with poorly designed optics that have significant aberration problems and are not very practical for photomicrography or serious quantitative microscopy.

Parfocal Distance - This is another specification that can often vary by manufacturer. Most companies produce objectives that have a 45 millimeter parfocal distance, which is designed to minimize refocusing when magnifications are changed.

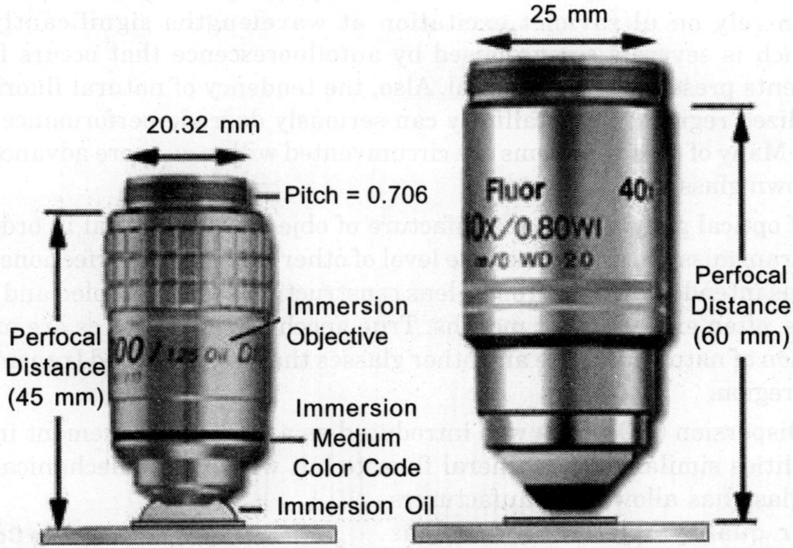

Fig. 3.14.

The objective depicted on the left in Fig. 3.14 has a parfocal distance of 45mm and is labelled with an immersion medium color code in addition to the magnification color code. Parfocal distance is measured from the nosepiece objective mounting hole to the point of focus on the specimen as illustrated in the figure. The objective on the right in Fig. 3.14 has a longer parfocal distance of 60 millimeters, which is the result of its being produced to the **Nikon CFI60 200/60/25 Specification**, again deviating from the practice of other manufacturers such as **Olympus** and **Zeiss**, who still produce objectives with a 45mm parfocal distance. Most manufacturers also make their objective nosepieces ***parcentric***, meaning that when a specimen is centreed in the field of view for one objective, it remains centreed when the nosepiece is rotated to bring another objective into use.

Glass Design - The quality of glass formulations has been paramount in the evolution of modern microscope optics, and there are currently several hundred of optical glasses available for the design of microscope objectives. The suitability of glass for the demanding optical performance of a microscope objective is a function of its physical properties such as refractive index, dispersion, light transmission, contaminant concentrations, residual autofluorescence, and overall homogeneity throughout the mixture. Care must be taken by optical designers to ensure that glass utilized in high-performance objectives has a high transmission in the near-ultraviolet region and also produces high extinction factors for applications such as polarized light or differential interference contrast.

Cements employed in building multiple lens elements usually have a thickness around 5-10 microns, which can be a source of artifacts in groups that have three or more lens elements cemented together. Doublets, triplets, and other multiple lens arrangements can display spurious absorption, transmission, and fluorescence characteristics that will disqualify the lenses for certain applications.

For many years, natural fluorite was commonly used in the manufacture of fluorite (semi-apochromat) and apochromat objectives. Unfortunately, many newly developed fluorescence techniques often rely on ultraviolet excitation at wavelengths significantly below 400 nanometers, which is severely compromised by autofluorescence that occurs from natural organic constituents present in this mineral. Also, the tendency of natural fluorite to exhibit widespread localized regions of crystallinity can seriously degrade performance in polarized light microscopy. Many of these problems are circumvented with new, more advanced materials, such as fluorocrown glass.

Annealing of optical glass for the manufacture of objectives is critical in order to remove stress, improve transmission, and reduce the level of other internal imperfections. Some of the glass formulations intended for apochromat lens construction are slow-cooled and annealed for extended periods, often exceeding six months. True apochromat objectives are manufactured with a combination of natural fluorite and other glasses that have reduced transmission in the near-ultraviolet region.

Extra Low Dispersion (**ED**) glass was introduced as a major advancement in lens design with optical qualities similar to the mineral fluorite but without its mechanical and optical demerits. This glass has allowed manufacturers to create higher quality objectives with lens elements that have superior optical corrections and performance. Because the chemical and optical properties of many glasses are of a proprietary nature, documentation is difficult or impossible to obtain. For this reason the literature is often vague about the specific properties of glasses utilized in the construction of microscope objectives.

Multilayer Antireflection Coatings - One of the most significant advances in objective design during recent years is the improvement in antireflection coating technology, which helps to reduce unwanted reflections (flare and ghosts) that occur when light passes through a lens system, and ensure high-contrast images. Each uncoated

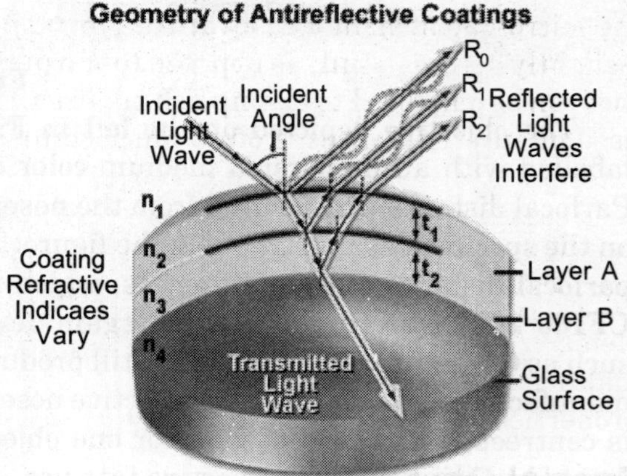

Fig. 3.15. Geometry of Antireflective Coatings.

air-glass interface can reflect between four and five percent of an incident light beam normal to the surface, resulting in a transmission value of 95-96 percent at normal incidence. Application of a quarter-wavelength thick antireflection coating having the appropriate refractive index can increase this value by three to four percent. As objectives become more sophisticated with an ever-increasing number of lens elements, the need to eliminate internal reflections grows correspondingly. Some modern objective lenses with a high degree of correction can contain as many as 15 lens elements having many air-glass interfaces. If the lenses were uncoated, the reflection losses of axial rays alone would drop transmittance values to around 50 percent. The single-layer lens coatings once utilized to reduce glare and improve transmission have now been supplanted by multilayer coatings that produce transmission values exceeding 99.9 percent in the visible spectral range. These specialized coatings are also used on the phase plates in phase contrast objectives to maximize contrast.

Illustrated in Fig. 3.15 is a schematic drawing of light waves reflecting and/or passing through a lens element coated with two antireflection layers. The incident wave strikes the first layer (**Layer A** in Fig. 3.15) at an angle, resulting in part of the light being reflected (**R(o)**) and part being transmitted through the first layer. Upon encountering the second antireflection layer (**Layer B**), another portion of the light is reflected at the same angle and interferes with light reflected from the first layer. Some of the remaining light waves continue on to the glass surface where they are again both reflected and transmitted. Light reflected from the glass surface interferes (both constructively and destructively) with light reflected from the antireflection layers. The refractive indices of the antireflection layers vary from that of the glass and the surrounding medium (air). As the light waves pass through the antireflection layers and glass surface, a majority of the light (depending upon the incident angle—usual normal to the lens in optical microscopy) is ultimately transmitted through the glass and focused to form an image.

Magnesium fluoride is one of many materials utilized in thin-layer optical antireflection coatings, but most microscope manufacturers now produce their own proprietary formulations. The general result is a dramatic improvement in contrast and transmission of visible wavelengths with a concurrent destructive interference in harmonically-related frequencies lying outside the transmission band. These specialized coatings can be easily damaged by mis-handling and the microscopist should be aware of this vulnerability. Multilayer antireflection coatings have a slightly greenish tint, as opposed to the purplish tint of single-layer coatings, an observation that can be employed to distinguish between coatings. The surface layer of antireflection coatings used on internal lenses is often much softer than corresponding coatings designed to protect external lens surfaces. Great care should be taken when cleaning optical surfaces that have been coated with thin films, especially if the microscope has been disassembled and the internal lens elements are subject to scrutiny.

From the discussion above it is apparent that objectives are the most important optical element of a compound microscope. It is for this reason that so much effort is invested in making sure that they are well-labeled and suited for the task at hand. We will explore other properties and aspects of microscope objectives in other sections of this tutorial.

Special Objectives

Perhaps as many as 90 percent of all optical microscopy investigations are conducted utilizing standard achromat or plan achromat objectives, which are the cheapest, most readily available, and already installed on a large base of microscopes around the world. A majority of microscope manufacturers also offer a wide variety of objectives having unique configurations designed to perform specific functions not normally found on common laboratory microscopes.

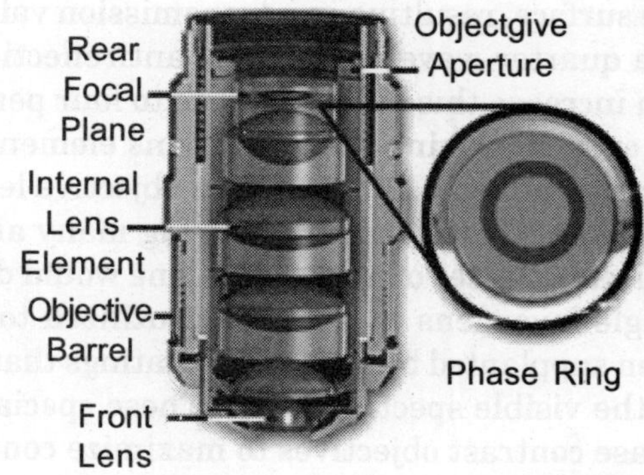

Fig. 3.16. Phase Contrast Objective.

Standard brightfield objectives, corrected for varying degrees of optical aberration, are the most common and are useful for examining specimens with traditional illumination techniques such as brightfield, darkfield, oblique, and Rheinberg. Several of these methods involve modifications to the substage condenser, but still utilize standard achromat, fluorite, and/or apochromat objectives either with or without flat-field correction. Other, more complex, techniques require specific objective configurations, which often include placement of a detector on or near the rear focal plane. To complicate the issue, the objective rear focal plane usually resides in the centre of an internal glass lens element, an area that is not easily accessible to the microscopist.

Objectives designed for phase contrast, Hoffman modulation contrast, and differential interference contrast require the assistance of optical detectors to modify events occurring at the objective rear focal plane. These objectives must be specially constructed to physically place the detecting element at the appropriate focal plane within the objective body. Phase contrast objectives require the insertion of a phase plate containing neutral density material and/or optical wave retarders at the rear focal plane. In addition, the phase plate must be placed conjugate to a matched annular ring positioned in the substage condenser. Hoffman objectives also require a modulation plate in the objective rear focal plane that is conjugate to a slit plate in the condenser. Differential interference contrast objectives generally do not require modification (other than the utilization of strain-free optical elements), but do rely on the action of Wollaston or Nomarski prisms strategically placed to influence optical path differences between sheared light beams at the rear focal plane. Other specialized objectives rely on modified optical elements, insertion of mirrors or oblique reflectors, adjustable apertures, and/or movable elements to perform unique functions. Details concerning the specific requirements of such objectives are discussed in the remainder of this section.

Phase Contrast

This is a classical method of introducing contrast into semi-transparent, unstained specimens such as microorganisms and cells in living tissue culture, which has been employed by biologists for the past 50 years. Phase contrast manipulates phase relationships between individual light rays as they are emitted from the specimen and translates them into amplitude or brightness changes that are visible to the microscopist. A special objective is required that is fitted with a darkened circular ring or groove (phase plate) fitted into the glass near the rear focal plane of the objective as illustrated in Figure 19. In addition, the condenser must also be modified with special annular openings suited to a particular magnification and objective. Phase contrast objectives are segregated into a number of categories depending upon the construction and neutral density of internal phase rings:

- **DL (Dark-Light)** - DL objectives produce a dark image outline on a light gray background. These objectives are designed to furnish the strongest dark contrast in specimens having major differences in refractive indices. The DL phase contrast objective is the most popular style for examination of cells and other semi-transparent living material and is especially suited for photomicrography and digital imaging.

- **DLL (Dark Lower Contrast)** - Similar to the DL objective, the DLL series allows better images in brightfield and is often used as a "universal" objective in microscope systems that utilize multiple illumination modes such as fluorescence, DIC, brightfield, and darkfield.
- **ADL (Apodized Dark-Light)** - Recently introduced by Nikon, the apodized phase contrast ADL objectives contain a secondary neutral density ring on either side of the phase ring. Addition of the secondary rings assists in reducing unwanted "halo" effects often associated with imaging in phase contrast microscopy.
- **DM (Dark-Medium)** - DM objectives produce a dark image outline on a medium gray background. These objectives are designed to be used for high image contrast with specimens having small phase differences, such as fine fibers, granules, and particles.
- **BM (Bright-Medium)** - Often referred to as negative phase contrast, BM objectives produce a bright image outline on a medium gray background. BM objectives are ideal for visual examination of bacterial flagella, fibrin bundles, minute globules, and blood cell counting.

In order to allow the microscopist to quickly identify phase contrast objectives, many manufacturers inscribe important specifications, such as the magnification, numerical aperture, tube length correction, etc., on the outer barrel in green letters. This serves to differentiate phase contrast objectives from ordinary brightfield, polarized, DIC, and fluorescence objectives which either use an alternative color code or the standard black lettering.

Differential Interference Contrast (DIC)

Nomarski differential interference contrast is also useful for unstained specimens, but is less effective with birefringent samples and can be used with reflected light for metallography and wafer inspection. DIC objectives are not modified internally, but are designed for use with

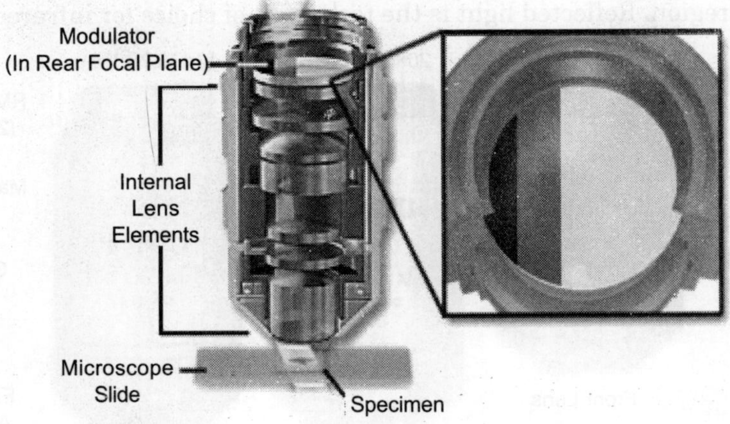

Fig. 3.17. Hoffman Modulation Contrast Objective

special magnification-dependent modified Wollaston or Nomarski prisms to produce high-contrast images. These objectives are also useful for brightfield, darkfield and other techniques when the prisms are removed from the optical path. Because DIC microscopy utilizes polarized light, strain must be minimized in objectives designed for this type of application. In the past, only strain-free achromats, planachromats, and some high-performance fluorite objectives were suitable for this task. However, recent improvements in lens design and antireflection coatings now allow utilization of apochromat objectives for DIC observation, photomicrography, and digital imaging. The barrel of objectives intended for use with a DIC prism is usually inscribed with the specific prism (low, medium, high, or 1, 2, 3, etc.) intended to be coupled with the objective.

Hoffman Modulation Contrast

Objectives for the Hoffman modulation contrast system are designed to increase visibility and contrast in unstained and living material by detecting optical gradients (or slopes) and converting them into variations of light intensity. Modulation contrast objectives have a unique optical amplitude spatial filter, termed a **modulator**, inserted into the rear focal plane (see Figure 3.17) of an achromat or planachromat objective (although a higher correction factor can also be utilized). The modulator has three zones of varying neutral density that transmit either one, fifteen, or one hundred percent of the light passing through the objective. Unlike the phase plate in phase contrast objectives, Hoffman modulators are designed not to alter the phase of light passing through any of the zones. When viewed under modulation contrast optics, transparent objects that are essentially invisible in ordinary brightfield microscopy take on an apparent three-dimensional appearance dictated by phase gradients in the specimen. Recent innovations in Hoffman objective design have yielded models that allow the contrast direction to be varied using a modulator inside the objective. Once adjusted, the contrast direction is maintained over the entire magnification range in a set of matched objectives.

Infrared Microscopy

Optical microscopy in the infrared region of the electromagnetic radiation spectrum is often undertaken for the study of materials that are uniformly transparent or opaque in the visible spectrum, but have significant absorption or transmission bands in the 700 nanometer-plus wavelength region. Reflected light is the technique of choice for infrared microscopy, and

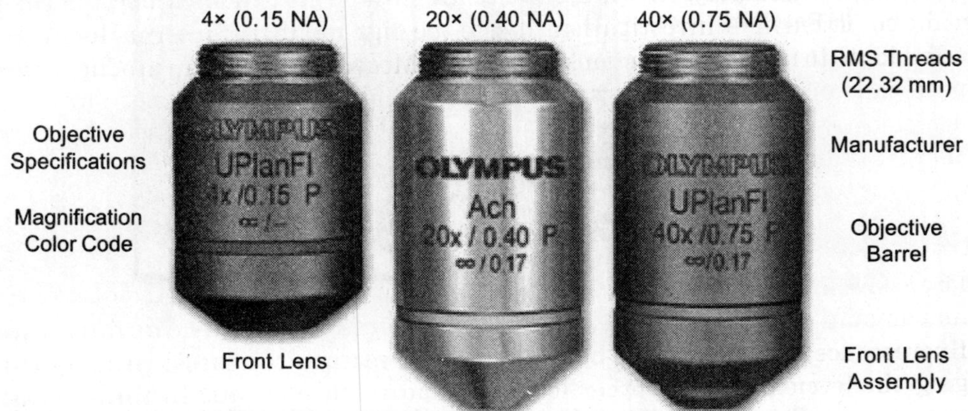

Fig. 3.18. Polarized Light Objective Identification Codes.

several specialized catoptric objectives have been designed to capture images with infrared light reflected from opaque specimens.

Although all microscope objectives transmit some degree of the shorter infrared wavelengths, very few are corrected for aberrations in this region and show a marked shift in focus when transcending from visible to infrared light. Most manufacturers offer specialized objectives with reduced numerical apertures that are designed to increase the depth of focus in specimens imaged with infrared light. Oil immersion objectives will not function normally with standard oils and the only suitable immersion fluid that is currently available for infrared microscopy is paraffin oil. The major concern with this type of microscopy is the ability to capture satisfactory photomicrographs using conventional imaging technology. Currently, several film emulsions are available that respond to the infrared spectrum, but electronic detectors are rapidly becoming the imaging device of choice for studies in this wavelength range.

Interference Microscopy

Interferometry is applied to the study of microscopic specimens by taking advantage of interference produced when light passing through an object is caused to interfere with a reference beam of light that has followed a somewhat different pathway. Under these circumstances, the opaque surface of a reflected light specimen or transparent specimens are imaged when the path difference between the two light beams is converted into intensity fluctuations. A wide variety of microscope and objective designs have been implemented for interference microscopy, many following the basic principles of the Michelson, Mach-Zehnder, and Jamin interferometers. Manufacturers of industrial microscopes (both reflected and transmitted illumination) often produce specialized objective/microscope combinations that take advantage of the optical interference phenomenon to achieve high-precision measurements.

Polarized Light

Unlike most other forms of microscopy, polarized light microscopy produces the best images when a minimum of optical elements are used in the construction of objectives. It is important to ensure that lens elements, optical cements, and antireflection coatings are free of both strain and birefringent materials that might interfere with quantitative assessment of specimen birefringence. Apochromatic objectives, the choice for most forms of microscopy, are generally not utilized for polarized light investigations due to the high number of internal lens elements that often contribute to internal reflections and strain. Most manufacturers produce objectives that are specifically optimized for polarized light and differential interference contrast, with fluorite-class objectives being used most often for these purposes. Objectives optimized for polarized light often have the exterior portion of the barrel painted black, with the specifications inscribed in bright red letters (Figure 3.21).

Ultraviolet Fluor Objectives

Epi-fluorescence applications require high numerical aperture objectives in order to capture the maximum amount of light from faintly emitting fluorescent specimens. The ratio of specific specimen fluorescence to the intrinsic background fluorescence becomes a primary concern when imaging single molecule and other low-light fluorescence events. In these situations, auto-fluorescence and/or internal reflection inside the objectives can interfere with imaging small structures and low fluorescence targets.

Fluorescence objectives are designed with quartz and special glasses that have high transmission from the ultraviolet (down to 340 nanometers) through the infrared regions of the electromagnetic radiation spectrum. These objectives are extremely low in auto-fluorescence in order to optimize the luminous flux emitted as secondary fluorescence by fluorophores attached to specimens. In addition to special lens elements, ultraviolet (UV) fluor objectives utilize specialized optical cements and antireflection coatings that are designed to operate through an extended range of fluorescence excitations throughout the spectrum. Correction for optical aberration and numerical aperture values in UV fluor objectives usually approaches that of apochromats, which contributes to image brightness and enhanced resolution of images generated by these advanced lenses (Figure 3.19). In addition, these objectives are designed with non-fluorescent glass to minimize artifacts arising from auto-fluorescence by internal lens elements. The primary drawback of high-performance fluorescence objectives is that many are not corrected for field curvature, resulting in images that do not have uniform focus throughout the view field. Although this problem is only of secondary concern when imaging weakly fluorescent specimens (especially with laser scanning confocal microscopy), it becomes a major issue when the objective is utilized to perform under conventional illumination techniques such as brightfield, darkfield, and differential interference contrast.

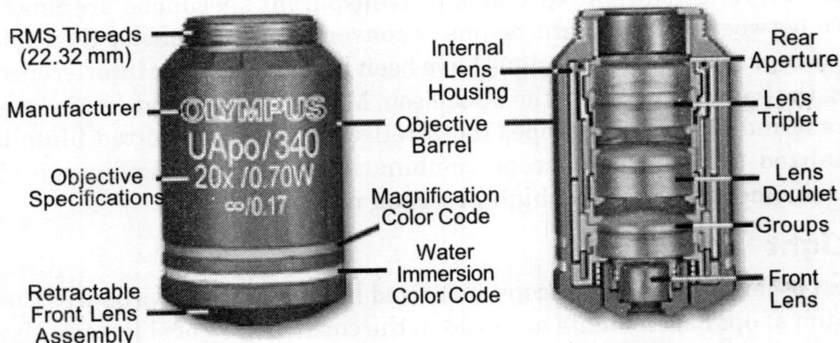

Fig. 3.19. Olympus Ultraviolet Fluorite Objective.

Reflected Light Objectives

Transmitted light objectives designed to be used with a cover glass are unsuitable for examining reflected light specimens whose surfaces are uncovered. Instead, specialized objectives that are corrected for observation and imaging of specimens without a cover glass are employed for reflected light microscopy. Today, most reflected light microscope objectives are infinity-corrected and are available in a broad spectrum of magnifications ranging from 5x to 200x. These objectives are manufactured in various qualities of chromatic and spherical correction, ranging from simple achromats to planachromats and planapochromats. Most, but not all, are designed to be used "dry" with air in the space between the objective and specimen. Many reflected light objectives are designed to focus at a longer working distance from the specimen than is usual (see below). Such objectives are labeled on the barrel of the objective as **LWD** (**L**ong **W**orking **D**istance), **ULWD** (Ultra-**L**ong **W**orking **D**istance), and **ELWD** (**E**xtra-**L**ong **W**orking **D**istance).

Objectives designed to be used with reflected darkfield illumination have a special construction consisting of a 360-degree hollow chamber surrounding the centrally located lens elements (Figure 3.20). Light from the illuminator passes through the periphery of the objective and is directed at the specimen from every azimuth in oblique rays to form a hollow cone of illumination. This is often accomplished by means of circular mirrors or prisms located at the bottom of the objective's hollow chamber. In this manner, the objective serves as two separate optical systems coupled coaxially such that the outer system functions as a "condenser" and the inner system as a typical objective.

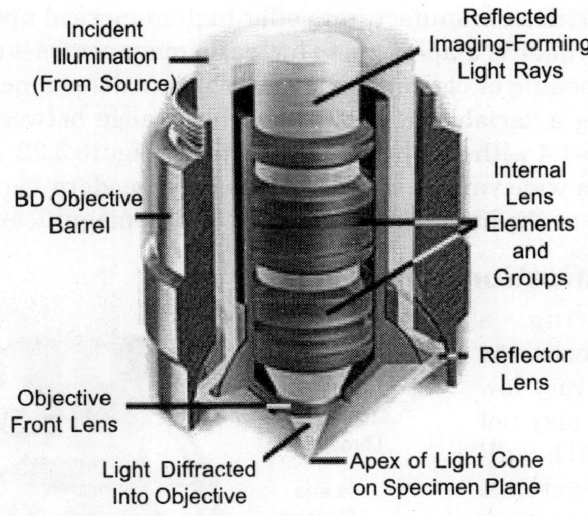

Fig. 3.20. Reflection Light Brightfield/Darkfield Objective.

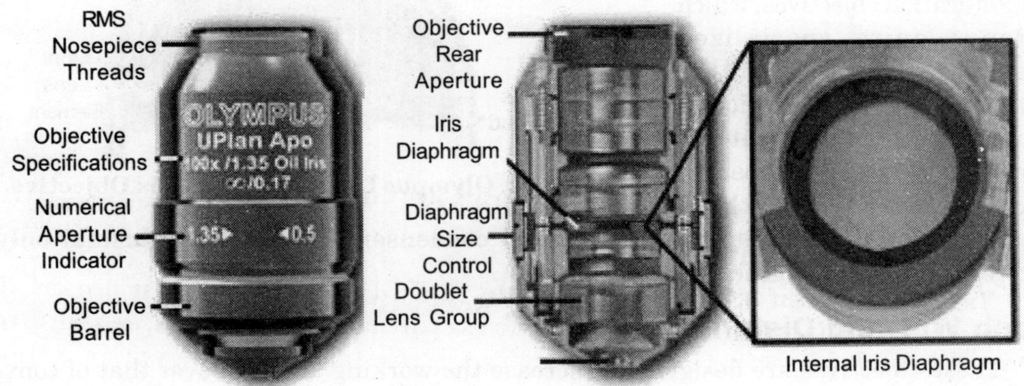

Fig. 3.21. Adjustable Numerical Aperture Objective.

The necessity of a hollow collar surrounding the lens elements in reflected light objectives requires that the diameter of the objective be significantly greater than that of ordinary brightfield objectives. In most cases, a nosepiece-mounting thread diameter larger than the Royal Microscopical Society (RMS) standard is used in reflected light objectives. This requires that reflected light darkfield objectives have a nosepiece with a larger thread size, which is typically referred to as a **BD** or **BF/DF** thread size. Most manufacturers offer objective adapters that convert standard RMS thread size nosepieces to BD thread size, allowing the use of these objectives on reflected light microscopes. Care should be taken to ensure that objectives used on BD thread nosepieces will conform to the tube length of the microscope.

Variable Numerical Aperture Objectives

Specimens with unusually high fluorescence quantum yields and/or very bright darkfield specimens often induce image flare by light emitted from areas outside the focal plane. To compensate for this artifact, manufacturers offer high numerical aperture objectives that are equipped with an internal iris diaphragm to increase image contrast during photomicrography or digital imaging. Opening or closing the iris diaphragm determines the size of the objective rear aperture yielding a variable numerical aperture range between 0.5 and the objective's upper limit (up to 1.35-1.4 with apochromatic objectives; Figure 3.22. Although iris diaphragms were once utilized in a wide variety of objective designs, modern variable numerical aperture objectives are usually at the high end (60x to 150x) of the magnification range.

Ultra-Low Magnification Objectives

Objectives having a magnification factor below 4x are considered to be very low in magnification and may not be compatible with all microscope optical systems. Generally, Köhler illumination is difficult to achieve with low magnification objectives, which often require specialized matching condensers to fill the rear aperture with sufficient illumination. Magnifications down to 0.5x have recently been achieved, but these objectives require special tube lenses and condensers, making them useful only on the microscopes for which they were designed.

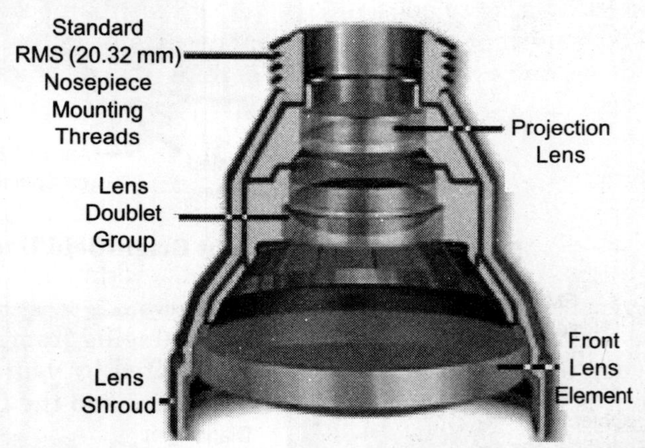

Fig. 3.22. Olympus Ultra-Low Mag 0.5× Objective.

Long Working Distance (LWD)

These objectives are designed to increase the working distance over that of conventional objectives through the use of special optical elements. The most useful application for LWD objectives is to view living cells in tissue culture through the walls of the thick vessels used to support the cells. Other uses for these objectives are to image samples through thick glass plates (for instance, between two microscope slides), or when micromanipulation of specimens must be conducted while viewing. Reflected light objectives are also produced with long working distances for examination of large specimens that would ordinarily be too bulky to fit within the confines of a normal microscope optical system.

Chapter 4 The Compound Microscope

Introduction

In modern times of science and technology, microscopes play a major role and that cannot be denied. Nowadays it is not only used specifically in few laboratories but can be found in schools, colleges and every other field wherein nature's creation are too small to be viewed by naked eyes and they have to be seen to understand the operations of the same. Therefore we need to understand how a microscope - a Compound Microscope works.

Compound means a microscope that has more than one lens and Microscope, if divided in to two different words as it is, "micro" means small and "scope" means view. Therefore in a whole the meaning is a device that would facilitate the viewing of small organisms with the help of multiple lenses. Simple Microscopes and Compound Microscope differ in the sense that Simple microscopes used a single lens and Compound Microscope uses more than one. Compound microscope can be used for several purposes ranging from medical research to a day at the beach as your urge to know can only be inhibited by your imagination. Like every field of knowledge it is first necessary to know the basics and then remember them as and how you keep working on it.

Basics of a Microscope

There are few basic functions that are same for every microscope. For every microscope to work there has to be a light source that would help to illuminate the object that has to be seen. The source of light can either be a mirror, which would reflect light from outside, or the device has to have a light source of its own. When light passes through the object, the lens that is nearest to the object will help to produce a magnified image of the specimen that is to be seen. The lens that is nearest to the object is called Objective Lens. The lens through which you look into the microscope is kid of a magnifying glass and helps to an extent to produce an enlarged image of that image that is already magnified by the objective lens of the specimen that is being viewed. So the magnification extent can be calculated by multiplying the power of *The Objective Lens and the Magnifier*. A Compound Microscope can magnify things up to 2000x, not more than that but then image larger than that neither our eyes would be able to recognize nor our brains would be able to read.

The Constitution of a Compound Microscope

A Compound light Microscope (See Fig. 4.1) has an eyepiece, stage clips which hold the slide in place, objectives that consist of different lenses, Adjustment knobs, power switch and a stage where the slides are kept and below it is at he source of light that is inbuilt in a Compound Light Microscope. There are other parts also like a Diaphragm, base, body tube; nose piece, aperture, arm and stage stop which helps to determine that height of the stage- all these parts facilitate the closer and clearer viewing of the object through the Compound Light Microscope. A Compound Microscope is nowadays used in several fields of sciences like the Microbiology, Botany, Geology, Genetics and the like. Forensic experts and scientists can also find out the country from which the drug has come by viewing its particles under a Compound Microscope as the shape of the crystals can give a reference as to which country the opium was grown in. Microscopes also help in looking at the minute details of the human cells and determine the presence or absence of minerals, identify the presence of metals, thus solve crimes and discover medicines.

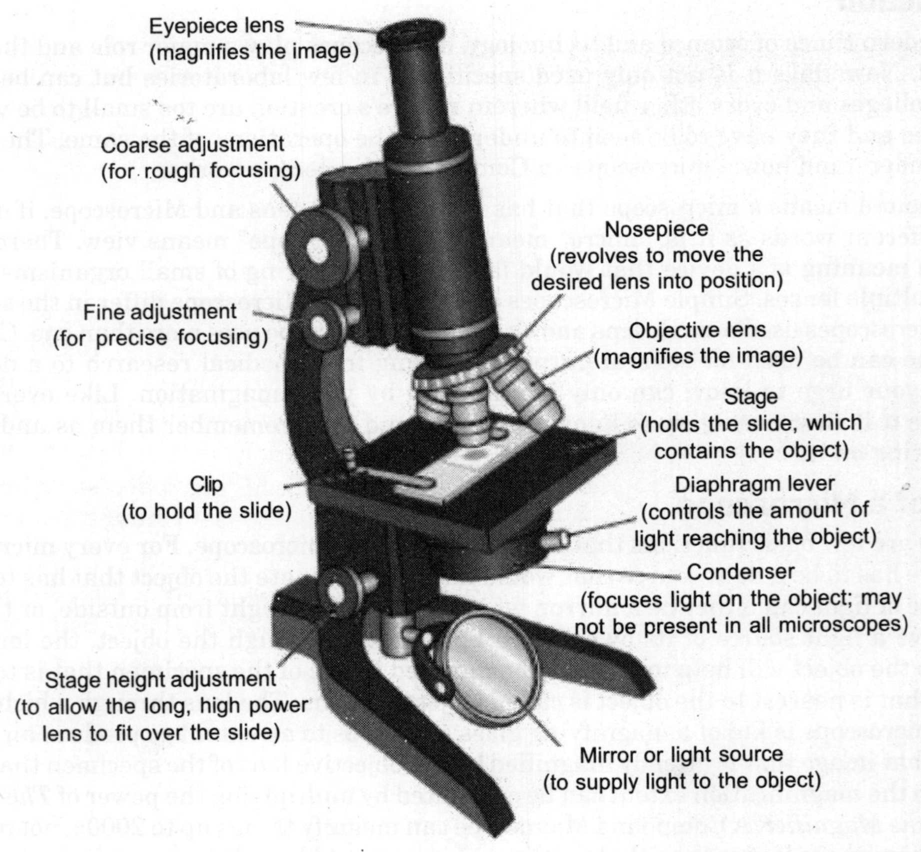

Fig. 4.1. The Elementary Compound Microscope and its different parts

Microscopes are used in schools and colleges, which help the children to look at the virus and bacteria that they learn about in textbooks. Things they could have missed pout while reading makes an impact when they are actually seeing the exact things in an enlarged form.

All optical microscopes share the same basic components:

- **The eyepiece** - a cylinder containing two or more lenses to bring the image to focus for the eye. The eyepiece is inserted into the top end of the body tube. Eyepieces are interchangeable and many different eyepieces can be inserted with different degrees of magnification. Typical magnification values for eyepieces include 5x, 10x and 2x. In some high performance microscopes, the optical configuration of the objective lens and eyepiece are matched to give the best possible optical performance. This occurs most commonly with apochromatic objectives.
- **The objective lens** - a cylinder containing one or more lenses, typically made of glass, to collect light from the sample. At the lower end of the microscope tube one or more objective lenses are screwed into a circular nose piece which may be rotated to select the required objective lens. Typical magnification values of objective lenses are 4x, 5x, 10x, 20x, 40x, 50x and 100x. Some high performance objective lenses may require matched eyepieces to deliver the best optical performance.
- **The stage** - a platform below the objective which supports the specimen being viewed. In the centre of the stage is a hole through which light passes to illuminate the specimen. The stage usually has arms to hold *slides* (rectangular glass plates with typical dimensions of 25 mm by 75 mm, on which the specimen is mounted).
- **The illumination source** - below the stage, light is provided and controlled in a variety of ways. At its simplest, daylight is directed via a mirror. Most microscopes, however, have their own controllable light source that is focused through an optical device called a condenser, with diaphragms and filters available to manage the quality and intensity of the light.

The whole of the optical assembly is attached to a rigid arm which in turn is attached to a robust U shaped foot to provide the necessary rigidity. The arm is usually able to pivot on its joint with the foot to allow the viewing angle to be adjusted. Mounted on the arm are controls for focusing, typically a large knurled wheel to adjust coarse focus, together with a smaller knurled wheel to control fine focus.

Updated microscopes may have many more features, including reflected light (incident) illumination, fluorescence microscopy, phase contrast microscopy and differential interference contrast microscopy, spectroscopy, automation, and digital imaging.

On a typical compound optical microscope, there are three objective lenses: a scanning lens (4×), low power lens (10×)and high power lens (ranging from 20 to 100×). Some microscopes have a fourth objective lens, called an oil immersion lens. To use this lens, a drop of immersion oil is placed on top of the cover slip, and the lens is very carefully lowered until the front objective element is immersed in the oil film. Such immersion lenses are designed so that the refractive index of the oil and of the cover slip is closely matched so that the light is transmitted from the specimen to the outer face of the objective lens with minimal refraction. An oil immersion lens usually has a magnification of 50 to 100×.

The actual power or magnification of an optical microscope is the product of the powers of the ocular (eyepiece), usually about 10×, and the objective lens being used.

Compound optical microscopes can produce a magnified image of a specimen up to 1000× and, at high magnifications, are used to study thin specimens as they have a very limited depth of field.

Operation

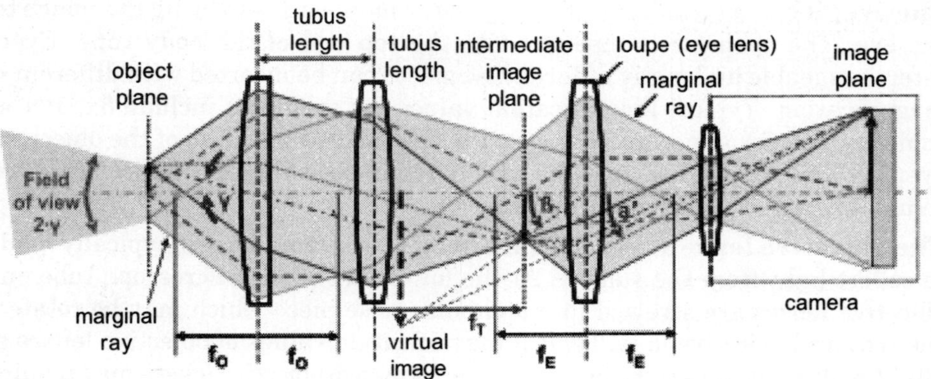

Fig. 4.2. The Optical path in a typical microscope

The optical components of a modern microscope are very complex and for a microscope to work well, the whole optical path has to be very accurately set up and controlled. Despite this, the basic optical principles of a microscope are quite simple (Fig. 4.2).

The objective lens is, at its simplest, a very high powered magnifying glass *i.e.* a lens with a very short focal length. This is brought very close to the specimen being examined so that the light from the specimen comes to a focus about 160 mm inside the microscope tube. This creates an enlarged image of the subject. This image is inverted and can be seen by removing the eyepiece and placing a piece of tracing paper over the end of the tube. By carefully focusing a brightly lit specimen, a highly enlarged image can be seen. It is this real image that is viewed by the eyepiece lens that provides further enlargement.

In most microscopes, the eyepiece is a compound lens, with one component lens near the front and one near the back of the eyepiece tube. This forms an air-separated couplet. In many designs, the virtual image comes to a focus between the two lenses of the eyepiece, the first lens bringing the real image to a focus and the second lens enabling the eye to focus on the virtual image.

In all microscopes the image is viewed with the eyes focused at infinity (mind that the position of the eye in the above figure is determined by the eye's focus). Headaches and tired eyes after using a microscope are usually signs that the eye is being forced to focus at a close distance rather than at infinity.

Advanced Compound Microscopes

Here we are going to discuss about the advanced compound microscopes, they are known as the binocular and the trinocular microscopes (These are used in most of the advanced research labs) (Fig. 4.3). It is to be noted that the basic mechanism of the binocular and trinocular microscope vision observation tubes are mostly same the only difference is the camera attachment for taking different photographs in the trinocular head. Advanced polarized light microscopes are often equipped with a Bertrand lens (sometimes referred to as an Amici-Bertrand lens) positioned on a movable sliding or tilting mount that is located between the analyzer and the eyepieces. In some cases, there is also a provision for focusing the Bertrand lens. When coupled to the eyepiece, the Bertrand lens provides a system that focuses on the objective rear focal plane, allowing the microscopist to observe illumination alignment, condenser aperture size, and conoscopic polarized light images. In Köhler illumination, an image of the lamp filament is formed in the objective rear focal plane, together with the image of the condenser aperture, so the Bertrand lens is often utilized to adjusting the illuminating (condenser) aperture diaphragm for optimum specimen contrast. The primary function in polarized light microscopy, however, is to view interference figures (conoscopic images). These images appear in the objective rear focal plane when an optically anisotropic specimen is viewed between crossed polarizers using a high numerical aperture objective/condenser combination.

Binocluar and tricocular Microscope Observation tube with Bertrand Lens

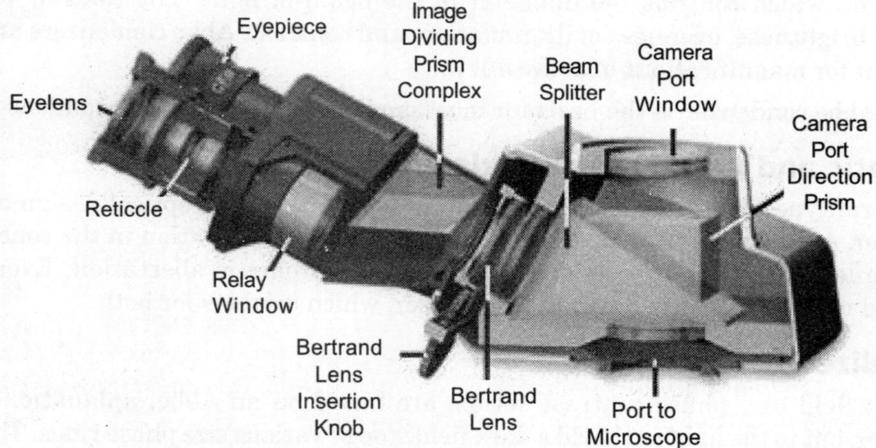

Fig. 4.3. The Mechanism of the Binocular and Trinocular Vision Observation Tube

Now we shall discuss about the sub stage parts of the microscopes that is the condensers, generally the basic and the advanced microscopes both have the condensers, but there are lots of differences in them, which we are going to discuss with illustration.

Condensers

A **condenser** is one of the main components of the optical system of many transmitted light compound microscopes. A condenser is a lens that serves to concentrate light from the illumination source that is in turn focused through the object and magnified by the objective lens. It is a basic component of almost all compound light microscopes manufactured since the 19th Century. An equivalent condenser, which focuses an electron beam, is a basic component of both transmission and scanning electron microscopes.

Early Condensers

Most condenser lens systems for research grade microscopes (1000X) have an N.A. value of 1.25. The 100X objective lens is also rated at 1.25. The medium between the 100X lens and the slide can be air or oil. Without getting too technical, the only way to get a Numerical Aperture greater than 1.0 is to use a material with a refractive index greater than 1.0. Oil (1.5) is such a material. So to get the best resolution at 1000X (N.A. 1.25), a drop of this special oil is placed between the lens and the slide. This is why most 100XR objective lenses are called oil immersion lenses. They work fine without oil but will give a sharper image with it (see image comparisons below). The oil can get messy and must be wiped up when you are finished. As a result, most people prefer to use their 100X objectives dry. If you try using immersion oil, clean up with lens tissue or a KimWipe on the lens. Avoid solvents.

Abbe Condenser

The Abbe condenser is named for its inventor, Ernst Karl Abbe, who developed it in 1870. The Abbe condenser is mounted below the stage of the microscope, and concentrates and controls the light that passes through the specimen and enters the objective. It has two controls, one which moves the Abbe condenser closer to or further from the stage, and another, the iris diaphragm, which controls the diameter of the beam of light. The controls can be used to optimize brightness, evenness of illumination, and contrast. Abbe condensers are particularly important for magnifications of above 400X.

The Abbe condenser is the basis for most modern light microscope condenser designs.

Aplanatic and Achromatic Condensers

Several types of condenser represent improvements on the optical design of a basic Abbe condenser. An aplanatic condenser corrects for spherical aberration in the concentrated light path, while an achromatic condenser corrects for chromatic aberration. Even more highly improved is an achromatic-aplanatic condenser, which corrects for both.

Specialized Condensers

Dark field and phase contrast setups are based on an Abbe, aplanatic, or achromatic condenser, but to the light path add a dark field stop or various size phase rings. These additional elements are housed in various ways. In most modern microscope (ca. 1990s–), such elements are housed in sliders that fit into a slot between the illuminator and the condenser lens. Many older microscopes house these elements in a turret-type condenser, these elements are housed in a turret below the condenser lens and rotated into place.In epifluorescence microscopy, the objective lens acts not only as a magnifier for the light emitted by the fluorescing object, but also as a condenser for the incident light.

Condensers and Numerical Aperture

Like objective lenses, condensers can vary in their numerical aperture (NA). Different condensers vary in their maximum and minimum numerical aperture, and a the numerical aperture of a single condenser varies depending on the where the diameter of the condenser aperture is set. In order for the maximum numerical aperture (and therefore resolution) of an objective lens to be realized, it must be matched with condenser set to like numerical aperture. The technique most commonly used in microscopy to optimize the light pathway between the condenser (and other illumination components of the microscope) and the objective lens is known as Köhler illumination. As with objective lenses, a condenser lens with a maximum numerical aperture of greater than 0.95 is designed to be used under oil immersion (or, more rarely, under water immersion), with a layer of immersion oil placed between the bottom of the slide and the top lens of the condenser. Without this oil layer, not only is maximum numerical aperture not realized, but the condenser is not able to precisely focus light on the object. Condensers with a numerical aperture of 0.95 or less are designed to be used without oil or other fluid on the top lens and are termed dry condensers. Dual dry/immersion condensers are basically oil immersion condensers that can nonetheless focus light with some degree of precision even without oil between the top lens and the slide. However, a dedicated dry condenser with an NA of 0.95 or less will focus light more precisely than a dry/immersion condenser used without oil immersion.

The substage condenser gathers light from the microscope light source and concentrates it into a cone of light that illuminates the specimen with uniform intensity over the entire viewfield. It is critical that the condenser light cone be properly adjusted to optimize the intensity and angle of light entering the objective front lens. Each time an objective is changed, a corresponding adjustment must be performed on the substage condenser to provide the proper light cone for the numerical aperture of the new objective.

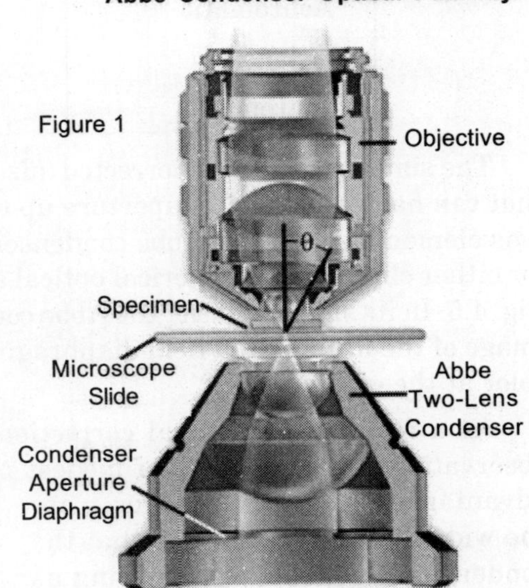

Fig. 4.4. The Function of an Advanced Substage Condenser

Aperture adjustment and proper focusing of the condenser are of critical importance in realizing the full potential of the objective. Specifically, appropriate use of the adjustable aperture iris diaphragm (incorporated into the condenser or just below it) is most important in securing correct illumination, contrast, and depth of field. The opening and closing of this iris diaphragm controls the angle of illuminating rays (and thus the aperture) which pass through the condenser, through the specimen and then into the objective. Visitors are invited to explore how changing the condenser aperture effects the illumination cone that explores the condenser numerical aperture. Condenser height is controlled by a rack and pinion gear system that allows the condenser focus to be adjusted for proper illumination of the specimen. Correct positioning of the condenser with relation to the cone of illumination and focus is critical to quantitative microscopy and optimum photomicrography.

Care must be taken to guarantee that the condenser aperture is opened to the correct position with respect to objective numerical aperture. When the condenser aperture diaphragm is opened too wide, stray light generated by refraction of oblique light rays from the specimen can cause glare and lower the overall contrast. On the other hand, when the aperture is made too small, the illumination cone is insufficient to provide adequate resolution and the image is distorted due to refraction and diffraction from the specimen.

Condensers are divided into classifications of purpose (e.g.: brightfield, darkfield, phase contrast, etc.), and also according to their degree of optical correction. There are four principle types of condensers with respect to correction of optical aberrations, as listed in Table 4.1.

Table 4.1. Condenser Aberration Corrections

Condenser Type	*Aberrations Coorected*	
	Spherical	*Chromatic*
Abbe	—	—
Aplanatic	x	—
Achromatic	—	x
Aplanatic-achromatic	x	x

The simplest and least corrected (also the least expensive) condenser is the Abbe condenser that can have a numerical aperture up to 1.4 in high-end models with three or more internal lens elements. While the Abbe condenser is capable of passing bright light, it is not corrected for either chromatic or spherical optical aberrations. A typical Abbe condenser is illustrated in Fig. 4.5. In its simplest form, the Abbe condenser has two optical lens elements that produce an image of the illuminated field diaphragm that is not sharp and is surrounded by blue and red color at the edges.

As a result of no optical correction, the Abbe condenser is suited mainly for routine observation with objectives of modest numerical aperture and magnification. The primary advantages of the Abbe condenser are the wide cone of illumination that the condenser is capable of producing as well as its ability to work with long working distance objectives. Most microscopes are supplied by the manufacturer with an Abbe condenser as the default and these condensers are real workhorses for routine laboratory use.

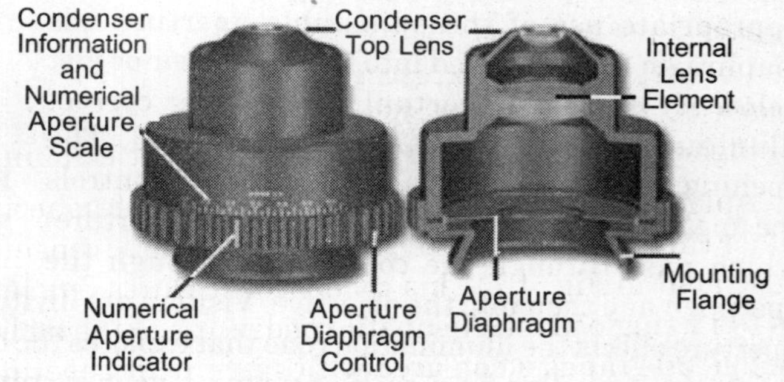

Fig. 4.5. Abbe Condenser (Numerical Aperture = 1.25)

The next level of condenser correction is split between the **aplanatic** and **achromatic** condensers that are corrected exclusively for either spherical (aplanatic) or chromatic (achromatic) optical aberrations. Typical examples of these condensers are illustrated in Fig. 4.6 (achromatic) and Fig. 4.7 (aplanatic). Achromatic condensers usually contain three to four lens elements and are corrected in two wavelengths (red and blue) for chromatic aberration.

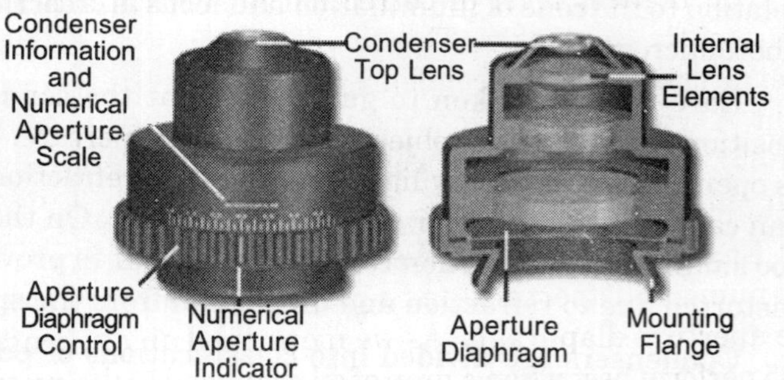

Fig. 4.6. Achromatic Condenser (Numerical Aperture = 0.95)

The achromatic condenser illustrated in Figure 4.6 contains four lens elements and has a numerical aperture of 0.95, the highest attainable without requiring immersion oil. This condenser is useful for both routine and critical laboratory analysis with "dry" objectives and also for black and white or color photomicrography.

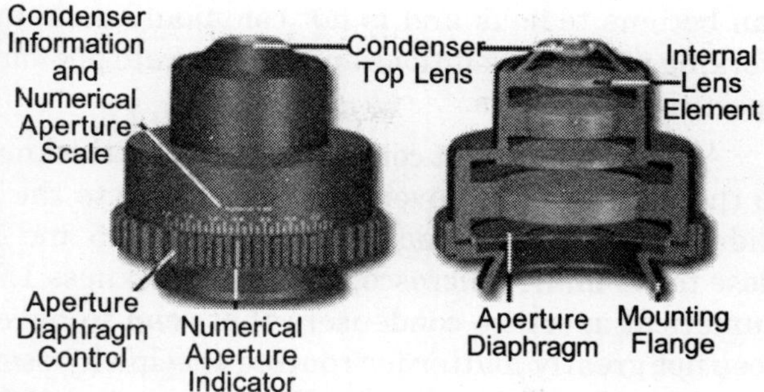

Fig. 4.7. Aplanatic Condenser (Numerical Aperture = 1.40)

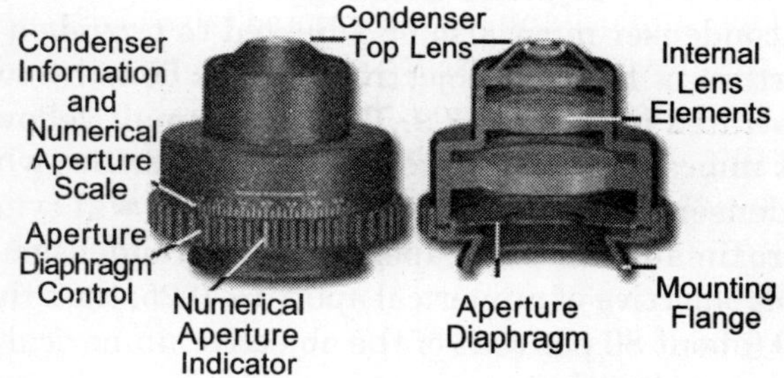

Fig. 4.8. Achromat/Aplanat Condenser (Numerical Aperture = 1.38)

A critical factor in choosing substage condensers is the numerical aperture performance that will be necessary to provide an illumination cone adequate for the objectives. The condenser numerical aperture should be equal to or slightly less than that of the highest objective numerical aperture. Therefore, if the highest magnification objective is an oil-immersion objective with a numerical aperture of 1.40, then the substage condenser should also have an equivalent numerical aperture to maintain the highest system resolution. In this case, immersion oil would have to be applied between the condenser top lens and the underside of the microscope slide to achieve the intended numerical aperture (1.40) and resolution. Failure to use oil will restrict the highest numerical aperture of the system to 1.0, the highest obtainable with air as the imaging medium.

Aplanatic condensers are well corrected for spherical aberration (green wavelengths) but not for chromatic aberration. A typical aplanatic condenser with a numerical aperture of 1.40 is illustrated in Fig. 4.7. This condenser features five lens elements and is capable of focusing light in a single plane. Aplanatic condensers are capable of producing excellent black and white photomicrographs when used with green light generated by either a laser source or by use of an interference filter with tungsten-halogen illumination.

The highest level of correction for optical aberration is incorporated in the aplanatic-achromatic condenser. This condenser is well corrected for both chromatic and spherical aberrations and is the condenser of choice for use in critical color photomicrography with white light. A typical aplanatic-achromatic condenser is illustrated in Figure 4.8(numerical aperture = 1.35). This condenser features eight internal lens elements cemented into two doublets and four single lenses.

Engravings found on the condenser housing include its type (achromatic, aplanatic, etc.), the numerical aperture, and a graded scale that indicates the approximate adjustment (size) of the aperture diaphragm. As we mentioned above, condensers with numerical apertures above 0.95 perform best when a drop of oil is applied to their upper lens in contact with the undersurface of the specimen slide. This ensures that oblique light rays emanating from the condenser are not reflected from underneath the slide, but are directed into the specimen. In practice, this can become tedious and is not commonly done in routine microscopy, but is essential when working at high resolutions and for accurate photomicrography using high-power (and numerical aperture) objectives.

Another important consideration is the thickness of the microscope slide, which is as crucial to the condenser as coverslip thickness is to the objective. Most commercial producers offer slides that range in thickness between 0.95 and 1.20 mm with the most common being very close to 1.0 mm. A microscope slide of thickness 1.20 mm is too thick to be used with most high numerical aperture condensers that tend to have a very short working distance. While this does not greatly matter for routine specimen observation, the results can be devastating with precision photomicrography. We recommend that microscope slides be chosen that have a thickness of 1.0 ± 0.05 mm, and that they be thoroughly cleaned prior to use.

When the objective is changed, for example from a 10X to 20X, the aperture diaphragm of the condenser must also be adjusted to provide a new light cone that matches the numerical aperture of the new objective. This is done by turning the knurled knob on the condensers illustrated in Figures 3.9. There is a small yellow arrow or index mark located on this knob that indicates the relative size of the aperture when compared to the linear gradation on the condenser housing. Many manufacturers will synchronize this gradation to correspond to the approximate numerical aperture of the condenser. For example, if the microscopist has selected a 10X objective of numerical aperture 0.25, then the arrow would be placed next the value 0.18-0.20 (about 80 per cent of the objective numerical aperture) on the gradation inscribed on the condenser housing.

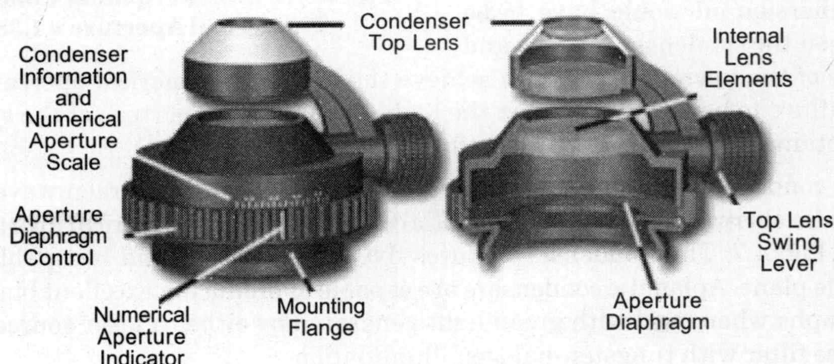

Fig. 4.9. Swing-Out Top Lens Condenser (Numerical Aperture = 1.35)

Often, it is not practical to use a single condenser with an entire range of objectives (2X to 100X) due to the broad range of light cones that must be produced to match objective numerical apertures. With low-power objectives in the range 2X to 5X, the illumination cone will have a diameter between 6-10 mm, while the high-power objectives (60X to 100X) need a highly focused light cone only about 0.2-0.4 mm in diameter. With a fixed focal length, it is difficult to achieve this wide range of illumination cones with a single condenser.

In practice, this problem can be solved in several ways. For low power objectives (below 10x), it may be necessary to unscrew the top lens of the condenser in order to fill the field of view with light. Some condensers are produced with a flip-top upper lens to accomplish this more readily, as illustrated in Figure 3.9. Many manufacturers now produce a condenser which flips over completely when used with low power objectives. Other companies may incorporate auxiliary correction lenses in the light path for securing proper illumination with objectives less than 10x. When the condenser is used without its top lens, the aperture iris diaphragm is opened wide and the field diaphragm, now visible at the back of the objective, serves as if it were the aperture diaphragm. Flip-top condensers are manufactured in a variety of configurations with numerical apertures ranging from 0.65 to 1.35. Those condensers that have a numerical aperture value of 0.95 or less are intended for use with "dry" objectives. However, flip-top condensers that have a numerical aperture greater than 0.95 are intended for use with oil-immersion objectives and they must have a drop of oil placed between the bottom of the microscope slide and the condenser top lens when examining critical samples.

In addition to the common brightfield condensers discussed above, there are a wide variety of specialized models suited to many different applications. Table 4.2 lists a number of different condenser configurations and the intended applications.

Table 4.2. Substage Condenser Applications

Condenser Type	*Brightfield*	*Darkfield*	*Phase Contrast*	*DIC*	*Polarizing*
Achromat/ Aplanat N.A. 1.3	• [10x~100x]				
Achromat Swing-out N.A. 0.90	• [4x~100x]				
Low-Power N.A. 0.20	• [1x~10x]				
Phase Contrast Abbe N.A. 1.25	•	• [up to N.A. 0.65]	• [10x~100x]		
Phase Contrast Achromat N.A. 0.85	•	• [up to N.A. 0.70]	• [4x~100x]		
DIC Universal Achromat/ Aplanat	•	• [up to N.A. 0.70]	• [10x, 100x]	• [20x, 40x, 100x]	
Darkfield, dry N.A. 0.80~0.95		• [4x~40x]			
Darkfield, oil N.A. 1.20~1.43		• [4x~100x]			
Stain-Free Achromat Swing-Out N.A. 0.90	•				• [4x~100x]

From the data in Table 4.2 it is obvious that substage condensers have a great deal of interchangeability among different applications. For instance, the DIC universal achromat/aplanat condenser is useful for brightfield, darkfield, and phase contrast, in addition to the primary DIC application. Other condensers have similar interchangeability.

Chapter

5

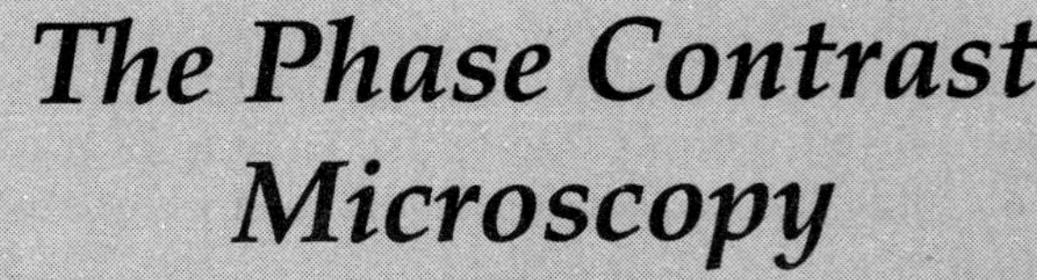

The Phase Contrast Microscopy

Introduction to Phase Contrast

The search was still on in the 1930's to find a way of using both direct and diffracted light from all azimuths to yield good contrast images of unstained objects that do not absorb light. Research by Frits Zernike during this period uncovered phase and amplitude differences between zeroth order and deviated light that can be altered to produce favorable conditions for interference and contrast enhancement.

Unstained specimens that do not absorb light are called phase objects because they slightly alter the phase of the light diffracted by the specimen, usually by retarding such light approximately 1/4 wavelength as compared to the undeviated direct light passing through or around the specimen unaffected. Unfortunately, our eyes as well as camera film, are unable to detect these phase differences. To reiterate, the human eye is sensitive only to the colors of the visible spectrum (variations in light frequency) or to differing levels of light intensity (variations in wave amplitude).

In phase specimens, the direct zeroth order light passes through or around the specimen undeviated. However, the light diffracted by the specimen is not reduced in amplitude as it is in a light-absorbing object, but is slowed by the specimen because of the specimen's refractive index or thickness (or both). This diffracted light, lagging behind by approximately 1/4 wavelength, arrives at the image plane **out of step** (also termed **out of phase**) with the undeviated light but, in interference, essentially undiminished in intensity. The result is that the image at the eyepiece level is so lacking in contrast as to make the details almost invisible.

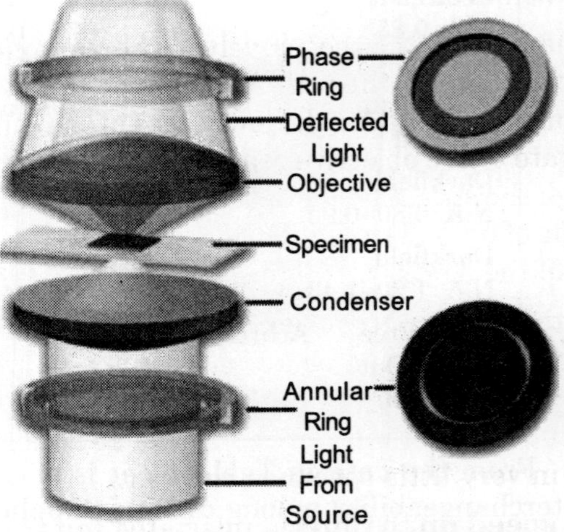

Fig. 5.1. Phase Contrast Light Pathways.

Zernike succeeded in devising a method—now known as **Phase Contrast** microscopy—for making unstained, phase objects yield contrast images as if they were amplitude objects. Amplitude objects show excellent contrast when the diffracted and direct light are out of step (display a phase difference) by 1/2 of a wavelength. Zernike's method was to **speed up** the direct light by 1/4 wavelength so that the difference in wavelength between the direct and deviated light for a phase specimen would now be 1/2 wavelength. As a result, the direct and diffracted light arriving at the image level of the eyepiece would be able to produce ***destructive interference*** (see the section on **image formation** for absorbing objects previously described). Such a procedure results in the details of the image appearing darker against a lighter background. This is called **dark** or **positive** phase contrast. A schematic illustration of the basic phase contrast microscope configuration is illustrated in Figure 5.1.

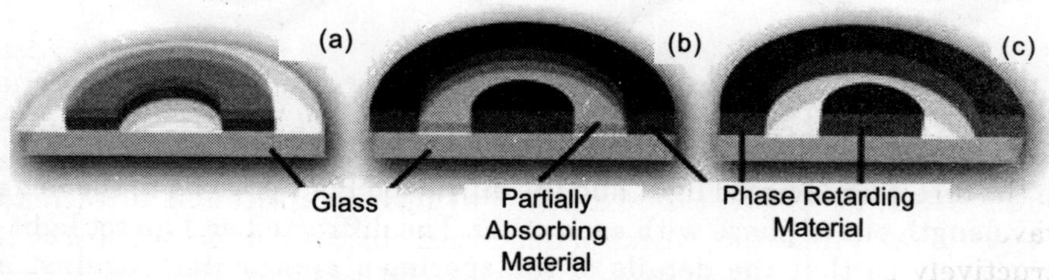

Fig. 5.2. Annular Phase Plates.

Another possible course, much less often used, is to arrange to **slow down** the direct light by 1/4 wavelength so that the diffracted light and the direct light arrive at the eyepiece in step and can interfere **constructively**. This arrangement results in a bright image of the details of the specimen on a darker background, and is called **negative** or **bright contrast**.

Phase contrast microscopy was very successful and ultimately gained widespread application, resulting in Zernike's award of the prestigeous Nobel prize in physics in 1953. The phase contrast technique has hailed as the greatest advance in microscopy in a century. Phase contrast, by "converting" phase specimens such as living material into amplitude specimens, allowed scientists to see details in unstained and/or living objects with a clarity and resolution never before achieved.

The Zernike method involves the **separation** of the direct zeroth order light from the diffracted light at the rear focal plane of the objective. To do this, a ring annulus is placed in position directly under the lower lens of the condenser at the front focal plane of the condenser, conjugate to the objective rear focal plane. As the hollow cone of light from the annulus passes through the specimen undeviated, it arrives at the rear focal plane of the objective in the shape of a ring of light. The fainter light diffracted by the specimen is spread over the entire rear focal plane of the objective.

If this combination were allowed, as is, to proceed to the image plane of the eyepiece, the diffracted light would be approximately 1/4 wavelength behind the direct light. At the image plane, the phase of the diffracted light would be out of phase with the direct light, but the amplitude of their interference would be almost the same as that of the direct light. This would result in very little specimen contrast.

To **speed up** the direct undeviated zeroth order light, a phase plate is installed with a ring shaped **phase shifter** attached to it at the rear focal plane of the objective. The narrow area of the phase plate is optically thinner than the rest of the plate. As a result, undeviated light passing through the phase ring travels a shorter distance in traversing the glass of the objective than does the diffracted light.

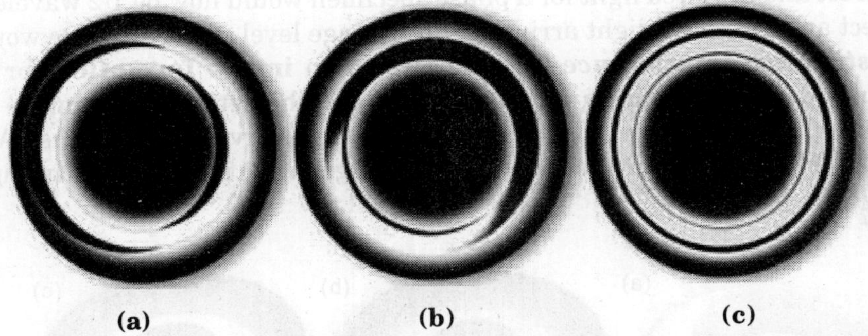

(a) (b) (c)

Fig. 5.3. Phase Rule and Light Annulus Alignments

Now, when the direct undeviated light and the diffracted light proceed to the image plane, they are 1/2 wavelength out of phase with each other. The diffracted and direct light can now interfere destructively so that the details of the specimen appear dark against a lighter background (just as they do for an absorbing or amplitude specimen). This is a description of what takes place in **positive** or **dark** phase contrast.

If the ring **phase-shifter** area of the phase plate were to be made thicker than the rest of the plate, direct light is slowed by 1/4 wavelength. In this case, the zeroth order light arrives at the image plane in step (or in phase) with the diffracted light, and constructive interference takes place. The image would appear bright on a darker background. The image appears bright on a darker background. This type of phase contrast is described as **negative** or **bright contrast**.

Because the undeviated light of the zeroth order is much brighter than the faint diffracted light, a thin absorptive transparent metallic layer is deposited on the ring to bring the direct and diffracted light into better balance of intensity in order to increase contrast. Also, because the speeding up or slowing down of the direct light is calculated on a 1/4 wavelength of green light, the phase image will appear best when a green filter is placed in the light path (a green interference filter is preferable). Such a green filter also helps achromatic objectives produce their best images, since achromats are spherically corrected for green light.

The accessories needed for phase contrast work are a substage phase contrast condenser equipped with annuli and a set of phase contrast objectives, each of which has a phase plate installed. The condenser usually has a brightfield position with an aperture diaphragm and a rotating turret of annuli (each phase objective of different magnification requires an annulus of increasing diameter as the magnification of the objective increases). Each phase objective has a darkened ring on its back lens. Such objectives can also be used for ordinary brightfield transmitted light work with only a slight reduction in image quality.

The phase outfit, as supplied by the manufacturer, usually includes a green filter and a phase telescope. The latter is used to enable the microscopist to alight the condenser annulus to **superimpose** it onto the ring of the phase plate. A set of centering screws in the substage condenser allows manipulation of the annulus to align it while observing the back focal plane of the objective with the phase telescope (or through a Bertrand lens).

To set up a phase microscope (cheek lining cells are a readily available test material), focus the specimen with the 10X phase objective. Next, configure the microscope for **Köhler illumination** using the brightfield (0) position of the condenser. This critical step is to assure the proper alignment of the microscope's objective, condenser, and field diaphragm. After the microscope is properly aligned, open up the condenser aperture diaphragm and swing the turret of the condenser into the 10 position (this usually automatically opens the aperture diaphragm). Place the green filter in the light path, and remove one of the eyepieces. Insert the phase telescope and, while observing the back of the objective, use the **Annulus Centering Screws** to center the annulus to the ring of the plate. A Bertrand lens or a pinhole eyepiece, if available, will allow a view of the back focal plane of the objective. Centering the phase annulus is often easier to do if the specimen is temporarily removed from the light path. After alignment of the phase ring with the annulus, reinsert the eyepiece and place the specimen back into its proper place on the microscope stage in the optical path.

Repeat the same procedure for each objective, making sure that the turret is rotated so the appropriate phase annulus is positioned to match the objective magnification. Some manufacturers provide individual push-in, centering, annuli that can be inserted into the lower part of the common Abbe condenser. Such inexpensive, simple devices do well with the 10X, 20X, and 40X phase objectives, but the condenser can receive only one at a time.

Phase microscopy continues to be a widely used and important tool, particularly for the microscopist studying living and/or unstained material, such as cells and tissues in culture. The method is also currently used simultaneously with reflected light fluorescence to reveal areas of a specimen that do not fluoresce. Phase microscopy techniques are particularly useful with specimens that are thin and scattered in the field of view.

There are some limitations of phase contrast microscopy:

- Phase images are usually surrounded by halos around the outlines of details. Such halos are optical artifacts, which sometimes obscure the boundaries of details.
- The phase annuli **do** limit the working numerical aperture of the optical system to a certain degree, thus reducing resolution.
- Phase contrast does not work well with thick specimens because of shifts in phase occur from areas slightly below or slightly above the plane that is in focus. Such phase shifts confuse the image and distort image detail.
- Phase images appear gray if white light is used and green if a green filter is used. In the past, many microscopists restricted their film to black and white when performing photomicrography on phase specimens. Today, many color films reproduce black, white, and grayscales very effectively, especially the tungsten-balanced transparency films from Fuji, Kodak, and Agfa.

Phase microscopy is another exemplification of how the manipulation of light at the substage condenser lower lens level and at the objective rear focal plane level has significant effect upon the image that is observed through the eyepiece.

Applications for Phase Contrast Microscopy

Phase contrast is preferable to bright field microscopy when high magnifications (400x, 1000x) are needed and the specimen is colorless or the details so fine that colour does not show up well.

Cilia and flagella, for example, are nearly invisible in bright field but show up in sharp contrast in phase contrast. Amoebae look like vague outlines in bright field, but show a great deal of detail in phase. Most living microscopic organisms are much more obvious in phase contrast.

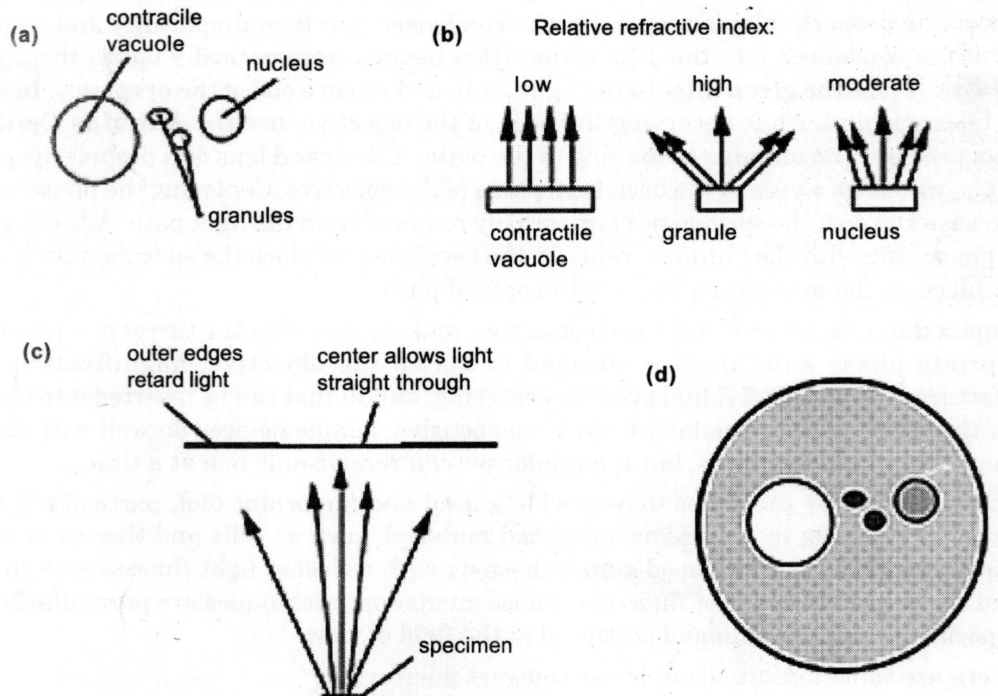

Fig. 5.4. Application of Phase Contrast (a) organelles are nearly invisible in bright field although they have different refractive indexes; (b) light is bent and retarded more by objects with a high refractive index; (c) in phase contrast a phase plate is placed in the light path. Barely refracted light passes through the center of the plate and is not retarded. Highly refracted light passes through the plate farther from center and is held back another one quarter wavelength.; (d) The microscope field shows a darker background (in this case the cell cytoplasm has a higher refractive index than the contractile vacuole), with the organelles in sharp contrast.

Using Phase Contrast

Phase contrast condensers and objective lenses add considerable cost to a microscope, and so phase contrast is often not used in teaching labs except perhaps in classes in the health professions and in some university undergraduate programs. This is unfortunate since the images obtainable in phase contrast mode can be very dramatic.

To use phase contrast the light path must be aligned. An element in the condenser is aligned with an element in a specialized phase contrast lens. This usually involves sliding a component into the light path or rotating a condenser turret. The elements are either lined up in a fixed position or are adjusted by the observer until the phase effect is optimized. Generally, more light is needed for phase contrast than for corresponding bright field viewing, since the technique is based on a diminishment of brightness of most objects.

Setting up your Microscope for Phase Contrast Observations

(you must have special phase objective lenses and a phase sub stage condenser)

Shown in Fig. 5.5 the phase contrast kit used on the National Optical 160 series microscopes. It consists of 4 objective lenses, a centering telescope and Zernike phase condenser lens.

The long adjustment screws on the phase condenser push in to engage set screws for proper alignment of the phase ring.

There is a thumb wheel on the opposite side used to dial in the proper setting to match the power of the objective lens. There is also a "BF" setting on the thumb wheel for brightfield. This allows you to use the phase objectives as standard brightfield lenses.

Not all phase contrast microscopes are the same but generally they rely on similar techniques to set up the system for optimum results. In the system shown at the left, the phase condenser has five settings that you spin with your thumb (10X, 20X, 40X, 100X and BF) BF is "brightfield", no phase.

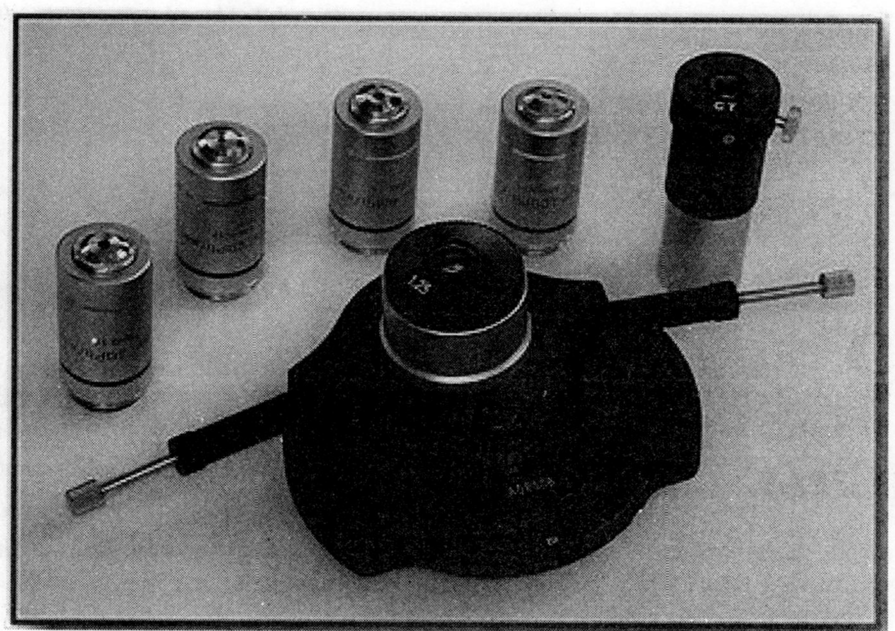

Fig. 5.5. The Full Phase Contrast Kit

To set up your microscope for phase optics, you first set it at BF and focus on the specimen. Adjust the height of the condenser for optimum image quality. Next, set the condenser turret to the phase setting for that particular lens and remove the specimen.

The controls that stick out from both sides on the back are for centering the condenser.

Next, you remove one of the eyepiece lenses and insert the centering telescope in its place(fig 5.7). The set screw is used to focus the centering telescope. When looking through the telescope, you will see two rings. They may or may not be concentric. By turning centering adjustment screws on the condenser, you align the rings so that they become concentric.

Fig. 5.6. Showing the Phase Contrast Kit Attached to a Microscope

Fig. 5.7. The phase Contrast Eye Piece.

Fig. 5.8A. The Allignment of the set up Phase Contrast Vision

Fig. 5.8B : The Alignment of the Phase Contrast set up

Fig. 5.8C. Author's Phase contrast Microscope

Figures 5.8A and 5.8B what you might see before alignment of the phase condenser

Once aligned and optimized for phase contrast microscopy You then remove the centering telescope and replace the eyepiece lens. Put your specimen back on the stage and you are ready for phase observations. When you change objectives, you should go through this procedure again (although you may discover that the alignment remains consistent with all objectives).

One final note. Your phase contrast microscope might also be a darkfield microscope. You will likely discover that when you set the condenser to a different phase setting than the chosen lens you will get a nice darkfield effect. So you are getting a pseudo darkfield microscope as well as a phase contrast instrument! This will show your specimen, not as a phase image, but illuminated on a black background.

Chapter 6

The Dark Field Microscopy

Dark field microscopy (dark ground microscopy) describes microscopy methods, in both light and electron microscopy, which exclude the unscattered beam from the image. As a result, the field around the specimen (i.e. where there is no specimen to scatter the beam) is generally dark. All of us are quite familiar with the appearance and visibility of stars on a dark night, this despite their enormous distances from the Earth. Stars can be readily observed at night primarily because of the stark contrast between their faint light and the black sky. Yet stars are shining both night and day, but they are invisible during the day because the overwhelming brightness of the sun "blots out" the faint light from the stars, rendering them invisible. During a total solar eclipse, the moon moves between the Earth and the sun blocking out the light of the sun and the stars can now be seen even though it is daytime. In short, the visibility of the faint star light is enormously enhanced against a dark background.

This principle is applied in darkfield (also called **darkground**) microscopy, a simple and popular method for making ***unstained*** transparent specimens clearly visible. Such objects often have refractive indices very close in value to that of their surroundings and are difficult to image in conventional brightfield microscopy. For instance, many small aquatic organisms have a refractive index ranging from 1.2 to 1.4, resulting in a negligible optical difference from the surrounding aqueous medium. These are ideal candidates for darkfield illumination.

Darkfield illumination requires blocking out of the central light which ordinarily passes through and around (surrounding) the specimen, allowing only oblique rays from every azimuth to "strike" the specimen mounted on the microscope slide. The top lens of a simple Abbe darkfield condenser is spherically concave, allowing light rays emerging from the surface in all azimuths to form an inverted hollow cone of light with an apex centered in the specimen plane. If no specimen is present and the numerical aperture of the condenser is greater than that of the objective, the oblique rays cross and all such rays will miss entering the objective because of their obliquity. The field of view will appear dark.

In optical microscopy, darkfield describes an illumination technique used to enhance the contrast in unstained samples. It works on the principle of illuminating the sample with light that will not be collected by the objective lens, so not form part of the image. This produces the classic appearance of a dark, almost black, background with bright objects on it. The darkfield condenser/objective pair illustrated in Fig. 6.1 is a high-numerical aperture arrangement that represents darkfield microscopy in its most sophisticated configuration, which will be discussed in detail below. The objective contains an internal iris diaphragm that serves to reduce the numerical aperture of the objective to a value below that of the inverted hollow light cone emitted by the condenser. The cardioid condenser is a reflecting darkfield design that relies on internal mirrors to project an aberration-free cone of light onto the specimen plane.

When a specimen is placed on the slide, especially an unstained, non-light absorbing specimen, the oblique rays cross the specimen and are diffracted, reflected, and/or refracted by optical discontinuities (such as the cell membrane, nucleus, and internal organelles) allowing these faint rays to enter the objective. The specimen can then be seen bright on an otherwise black background. In terms of Fourier optics, darkfield illumination removes the zeroth order (unscattered light) from the diffraction pattern formed at the rear focal plane of the objective. This results in an image formed exclusively from higher order diffraction intensities scattered by the specimen (See fig. 6.3)

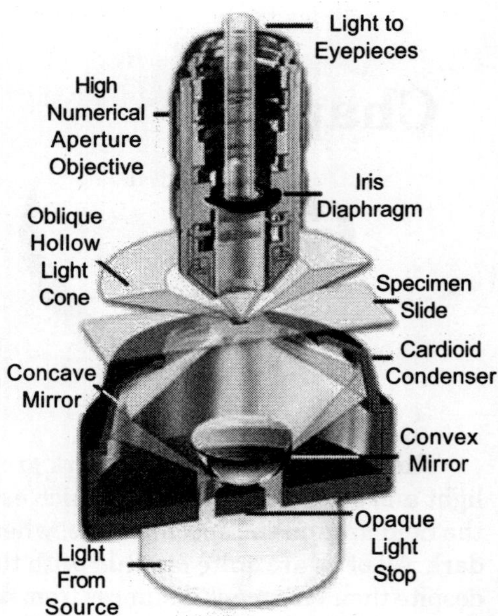

Fig. 6.1. Cardioid Darkfield Condenser.

The Light's Path

The steps are illustrated in Fig. 6.2 where an upright microscope is used:

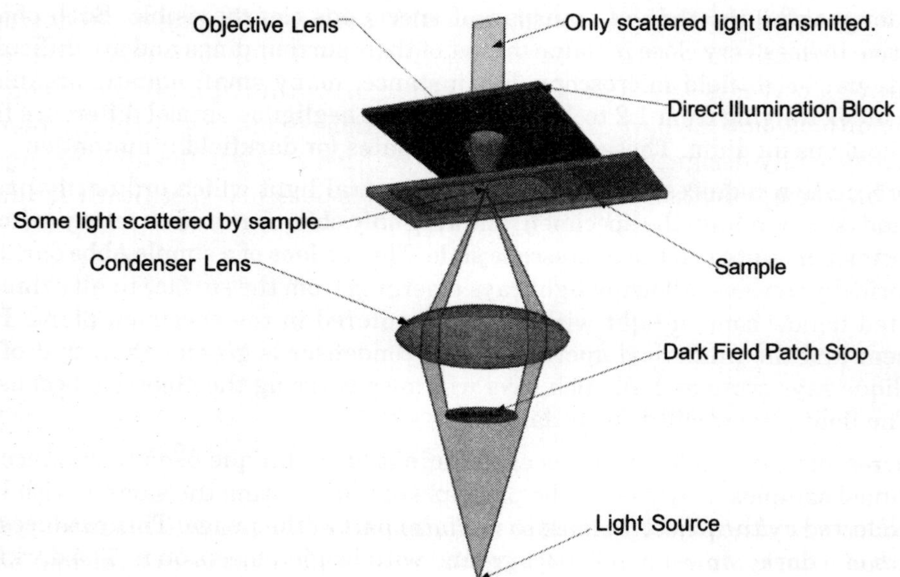

Fig. 6.2. The Diagram illustrating the light path through a dark field microscope.

1. Light enters the microscope for illumination of the sample.
2. A specially sized disc, the *patch stop* (see figure) blocks some light from the light source, leaving an outer ring of illumination.
3. The condenser lens focuses the light towards the sample.
4. The light enters the sample. Most is directly transmitted, while some is scattered from the sample.
5. The **scattered light** enters the objective lens, while the **directly transmitted light** simply misses the lens and is not collected due to a *direct illumination block* (see Fig. 6.2).
6. Only the scattered light goes on to produce the image, while the directly transmitted light is omitted.

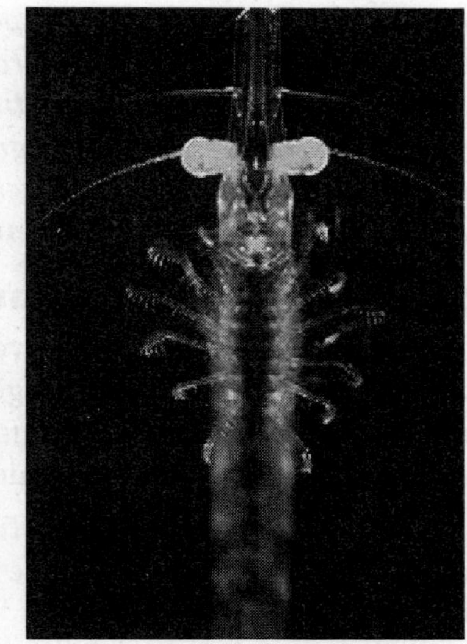

Fig. 6.3. Dark field microscopy produces an image with a dark background.

This is the Only Common Factor

Specimens that have smooth reflective surfaces produce images due, in part, to **reflection** of light into the objective. In situations where the refractive index is different from the surrounding medium or where refractive index gradients occur (as in the edge of a membrane), light is **refracted** by the specimen. Both instances of reflection and refraction produce relatively small angular changes in the direction of light, allowing some to enter the objective. In contrast, some light striking the specimen is also **diffracted**, producing a 180-degree arc of light that passes through the entire numerical aperture range of the objective. The resolving power of the objective is the same in darkfield illumination as observed under bright-field conditions, but the optical character of the image is not as faithfully reproduced (except when a specially-designed iris diaphragm is utilized to lower the effective numerical aperture with high-magnification oil immersion objectives designed specifically for darkfield microscopy).

As in the example of starlight described above, the visibility is greatly enhanced by the contrast between the brightly shining specimen and the dark surround. As discussed above, what has happened in darkfield illumination is that all the ordinarily undeviated rays of the zeroth order have been blocked by the opaque stop. Oblique rays, now diffracted by the specimen and yielding first, second, and higher diffracted orders at the rear focal plane of the objective, proceed onto the image plane where they interfere with one another to produce an image of the specimen.

Ideal candidates for dark field illumination include minute living aquatic organisms, diatoms, small insects, bone, fibers, hair, unstained bacteria, yeast, cells in tissue culture, and protozoa. Non-biological specimens include mineral and chemical crystals, colloidal particles, dust-count specimens, and thin sections of polymers and ceramics containing small inclusions, porosity differences, or refractive index gradients. Care should be taken when preparing specimens for darkfield microscopy because features that lie above and below the plane of focus can also scatter light and contribute to image degradation. Specimen thickness and microscope slide thickness are also very important and, in general, a thin specimen is desirable to eliminate the possibility of diffraction artifacts that can interfere with image formation.

Advantages and Disadvantages

Dark field microscopy is a very simple yet effective technique and well suited for uses involving live and unstained biological samples, such as a smear from a tissue culture or individual water-borne single-celled organisms. Considering the simplicity of the setup, the quality of images obtained from this technique is impressive.

The main limitation of dark field microscopy is the low light levels seen in the final image. This means the sample must be very strongly illuminated, which can cause damage to the sample.

Dark field microscopy techniques are almost entirely free of artifacts, due to the nature of the process. However the interpretation of dark field images must be done with great care as common dark features of bright field microscopy images may be invisible, and vice versa.

While the dark field image may first appear to be a negative of the bright field image, different effects are visible in each. In bright field microscopy, features are visible where either a shadow is cast on the surface by the incident light, or a part of the surface is less reflective, possibly by the presence of pits or scratches. Raised features that are too smooth to cast shadows will not appear in bright field images, but the light that reflects off the sides of the feature will be visible in the dark field images.

Conventional Darkfield Imaging

Briefly, conventional darkfield imaging involves tilting the incident illumination until a diffracted, rather than the incident, beam passes through a small objective aperture in the objective lens back focal plane. Darkfield images, under these conditions, allow one to map the diffracted intensity coming from a single collection of diffracting planes as a function of projected position on the specimen, and as a function of specimen tilt.

In single crystal specimens, single-reflection darkfield images of a specimen tilted just off the Bragg condition allow one to "light up" only those lattice defects, like dislocations or precipitates, which bend a single set of lattice planes in their neighborhood. Analysis of intensities in such images may then be used to estimate the amount of that bending. In polycrystalline specimens, on the other hand, darkfield images serve to light up only that subset of crystals which is Bragg reflecting at a given orientation.

The substage condensers illustrated in Fig. 6.4 demonstrate the effect of an opaque stop on light pathways through a simple refracting condenser. On the left (Fig. 6.4(a)), is a typical Abbe brightfield condenser positioned with the aperture diaphragm opened to maximize the numerical aperture of the light cone. Light from the source passes through the aperture diaphragm and is then refracted through several lens elements to form an inverted cone of light with a numerical aperture of approximately 1.20. When an opaque spider-style light stop (Fig. 6.4 (b)) is inserted below the completely opened aperture diaphragm, the central light rays are blocked, allowing only peripheral light rays to pass through the lenses to form an inverted oblique hollow cone of light with no change in numerical aperture (1.20). The illuminating hollow light cone is formed by refraction of light at perimeter of the lens elements, where optical correction is often poorest. Even so, this condenser will perform adequately using low magnification objectives, and produces very nice results for qualitative darkfield work. For more exacting quantitative microscopy, it is necessary to use aplanatic condensers (corrected for both chromatic and spherical aberrations), which perform much better by producing images with clearer detail and more reliable features.

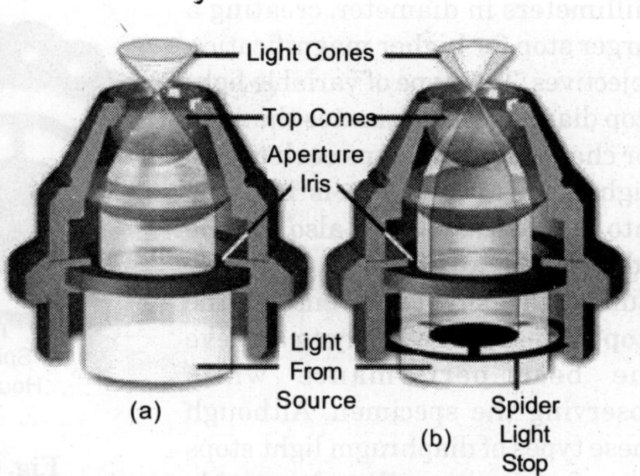

Fig. 6.4. Abbe Darkfield Condenser.

In a darkfield microscope, if you were to look at the back of the objective through a Bertrand lens or phase telescope, it would appear filled with light. This faint diffracted light is reconstituted into the visible image at the plane of the eyepiece diaphragm with its contrast reversed, bright image on black background. Since darkfield microscopy eliminates the bright undeviated light, this form of illumination is very wasteful of light and thus demands a high intensity illumination source. Microscope slides must be of the appropriate thickness, approximately one millimeter +/- 0.1 mm. Specimen slides and all optical surfaces in the light path must be scrupulously clean because every dirt speck will be mercilessly bright.

There are several pieces of equipment that are utilized to produce darkfield illumination. The simplest is a "spider stop" placed just under the bottom lens (in the front focal plane) of the substage condenser. Both the aperture and field diaphragms are opened wide to pass oblique rays. The central opaque stop (you can make one by mounting a coin on a clear glass disk) blocks out the central rays. This device works fairly well, even with the Abbe condenser, with the 10x objective up to 40x or higher objectives having a numerical aperture no higher than 0.65. The diameter of the opaque stop should be approximately 16-18 millimeters for a 10x objective of numerical aperture 0.25 to approximately 20-24 millimeters for 20x and 40x objectives of numerical apertures approaching 0.65.

The set of stops illustrated in Figure 6.7(a) vary in size from 8 millimeters to 30 millimeters and provide excellent darkfield opaque stops for virtually any numerical aperture objective (below 0.65). Individual stops can be interchanged simply by removing the screw in the bottom of the support spider and replacing the stop with a new size. The outer size of the spider holder will vary depending upon the housing opening diameter at the bottom of the condenser.

The light stop illustrated in Figure 6.7(b) is an ingenious device that expands and contracts a "reverse iris" diaphragm to increase or decrease the size of the stop using a lever control arm. As this lever is turned, the size of the central leaves changes from about 10 millimeters to 25 millimeters in diameter, creating a larger stop for higher magnification objectives. This type of variable light stop diaphragm eliminates the need for changing light stops each time a higher power objective is inserted into the optical path. It also has the additional advantage of being "tunable" for slight differences in the stop diameter necessary to achieve the best performance while observing the specimen. Although these types of diaphragm light stops are now very rare, they do provide a unique method of achieving highly desirable effects with darkfield illumination.

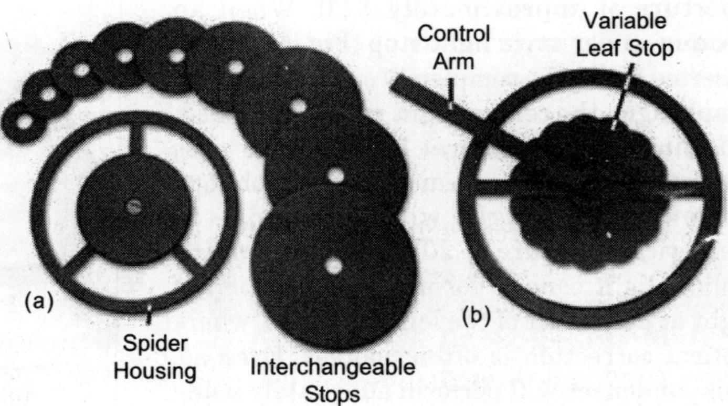

Fig. 6.5. Stops for Darkfield Illumination.

Almost any brightfield laboratory microscope can be easily converted for use with darkfield illumination. As discussed above, central opaque stops can be fashioned from a small coin, cardboard, plastic, or black paper that can be placed in a filter carrier beneath the condenser (or taped to the condenser bottom with adhesive tape) to block light from entering the front lens of the objective. The diameter of the opaque stop will vary from objective to objective and should be carefully measured by placing a transparent ruler in the substage filter carrier and holding it steady against the bottom of the condenser. Next, determine the opening size by removing an eyepiece and observing the image of the ruler at the back focal plane of the objective using a phase telescope (or by inserting a Bertrand lens). Make certain that both the substage condenser aperture and field diaphragms are opened to their widest position before performing this maneuver. The number of ruler divisions visible in the back focal plane will be equal to the size of the stop necessary to block zeroth order light from entering the objective. Change to the next largest size objective and take another measurement, repeating until stop sizes for all objectives are known.

A guide to approximate opaque stop size versus magnification is given in Table 6.1.

Table 6.1. Approximate Field Stop Diameter Size.

Magnification	*Numerical Aperture*	*Stop Size (mm)*
1X	0.03	25-30
2x	0.05	8-11
4X	0.10	8-14
10x	0.25	16-18
20X	0.40	18-20
20x	0.65	20-22
40X	0.65	22-24

The actual size will vary, depending upon several factors including the proximity of the stop with respect to the condenser aperture diaphragm, the numerical aperture of both the objective and condenser, the degree of aberration correction for the condenser, and the field number of the eyepiece. Also important in determining the stop size is the diameter of the condenser back lens, the magnification power of the eyepiece (smaller magnifications require slightly larger stops), and the type of mounting medium. Stop size varies proportionally to the refractive index of the mounting medium: higher refractive index requires a larger stop. A dry mount will also need a smaller stop than an aqueous suspension.

Use scissors or (preferably) a brass cork borer to cut a set of stops matched to all of the objectives, and glue them to a sturdy sheet of clear acetate or glass. The acetate or glass substrate should be easily mountable onto the underside of the substage condenser, either through a filter holder or by other means, such as adhesive tape. Alignment of the stop can be done by observing it through a Bertrand lens or removing the eyepiece and viewing through a phase telescope while adjusting the condenser centering screws.

Darkfield Microscopy at High Magnifications

For more precise work and blacker backgrounds, you may choose a condenser designed especially for darkfield, i.e. to transmit only oblique rays. There are several varieties: "dry" darkfield condensers with air between the top of the condenser and the underside of the slide—and immersion darkfield condensers which require the use of a drop of immersion oil (some are designed to use water instead) establishing contact between the top of the condenser and the ***underside*** of the specimen slide. The immersion darkfield condenser has internal mirrored surfaces and passes rays of great obliquity and free of chromatic aberration, producing the best results and blackest background.

Perhaps the most widely used darkfield condenser is the paraboloid, consisting of a solid piece of glass ground very accurately into the shape of a paraboloid, as illustrated in Fig. 6.6(b). Light incident upon the reflecting surface (between the glass and condenser housing in Fig. 6.6(b)) of a paraboloid condenser will be focused at the focal point of the reflector. Most paraboloid condensers are cut to ensure that the focal point is slightly beyond the top of the condenser so that parallel light rays will be focused at a position that maximizes illumination of the specimen. The light stop at the bottom of the glass condenser serves to block central rays from reaching the specimen. Light rays that are reflected by the condenser are angled higher than the critical angle of reflection and converge at the principal focus of the condenser. The combination of a glass slide, mounting medium, and immersion oil (between the condenser and the microscope slide) complete the optical homogeneity of the paraboloid shape.

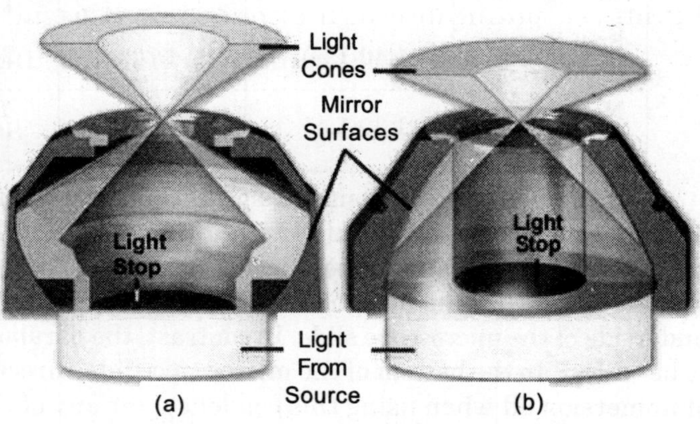

Fig. 6.6. Reflecting High Numerical Aperture Darkfield Condensers.

As discussed above, the dry darkfield condenser is useful for objectives with numerical apertures below 0.75, while the paraboloid and cardioid immersion condensers can be used with objectives of very high numerical aperture (up to 1.4). Objectives with a numerical aperture above 1.2 will require some reduction of their working aperture since their maximum numerical aperture may exceed the numerical aperture of the condenser, thus allowing direct light to enter the objective. For this reason, many high numerical aperture objectives designed for use with darkfield as well as brightfield illumination are made with a built-in adjustable iris diaphragm that acts as an aperture stop. This reduction in numerical aperture also limits the resolving power of the objective as well as the intensity of light in the image. Specialized objectives designed exclusively for darkfield work are produced with a maximum numerical aperture close to the lower limit of the numerical aperture of the darkfield condenser. They do not have internal iris diaphragms, however the lens mount diameters are adjusted so at least one internal lens has the optimum diameter to perform as an aperture stop.

Table 6.2 lists several properties of the most common reflecting high numerical aperture darkfield condensers. This table should be used as a guide when selecting condenser/objective combinations for use with high numerical aperture darkfield applications.

Table 6.2. High Numerical Aperture Darkfield Condenser Specifications

Condenser Type	*Hollow Cone Numerical Aperture*	*Objective Maximum Numerical Aperture*	*Number of Reflecting Surfaces*	*Optical Corrections*
Paraboloid	1.00-1.40	0.85	1 Parabolic	Achromatic
Cardioid	1.20-1.30	1.05	1 Spherical 1 Cardioidal	Achromatic/Aplanatic
Bicentric	1.20-1.30	1.05	1 Cardioidal Spherical	Aplanatic
Bispheric	1.20-1.30	1.05	2 Spherical	Aplanatic
Cassegrain	1.40-1.50	1.30	1 Aspheric 1 Spherical	Aplanatic
Spot Ring (Bicentric)	1.40-1.50	1.30	2 Spherical	Aplanatic
Nelson Cassegrain	1.30-1.45	1.20	1 Aspheric 1 Spherical	Aplanatic

The condensers are designed specifically to produce oblique hollow light cones of high numerical aperture for darkfield illumination. In both instances, the upper surface of the condenser is planar and perpendicular to the optical axis of the microscope. The condenser on the left (Fig. 6.6(a)) is designed to be used "**dry**" with no oil between the condenser and the underside of the microscope slide. In contrast, the paraboloid condenser in Fig. 6.6(b) is intended to be "**oiled**" to the bottom of the microscope slide, directly underneath the specimen. Omission of immersion oil when using this condenser (or any of the other condensers listed in Table 6.2 will preclude any light from reaching the specimen. The oblique hollow cone of light rays emitted by these condensers cannot emerge from the top lens without oil and will be totally reflected back into the condenser. Light emitted from the illumination source is reflected at the mirrored glass surfaces within the interior of the condensers and exits the top of the condensers at much higher angles of inclination than the critical angle (approximately 41 degrees) at which total reflection occurs for passage of light from glass to air. In the situation of the oiled paraboloid condenser (Figure 6.6(b) and the condensers in Table 6.2) where the refractive index of the condenser glass, immersion oil, and glass slide are equal, light emitted from the condenser passes through the specimen unrefracted by glass-air interfaces.

Reflecting high numerical aperture condensers listed in Table 6.2 cover a wide range of designs used to produce the oblique hollow cone of light necessary for high-magnification darkfield microscopy. The paraboloid darkfield condenser has been discussed in detail above. Another very useful design is the cardioid condenser that is illustrated in Fig. 6.1. This condenser design utilizes a mirrored hemisphere in the center of the condenser that serves as both a light stop and a reflector to direct light onto a second reflecting surface shaped to resemble a cardioid of revolution, from which the condenser derives its name. The combination of spherical and cardioid reflecting surfaces produces a condenser that is free from coma and both spherical and chromatic aberration. There are several technical drawbacks to using a condenser of such high numerical aperture. The cardioid condenser is very sensitive to alignment and must be carefully positioned to take advantage of the very sharp cone of illumination, making it the most difficult darkfield condenser to use. In addition, the condenser produces a significant amount of glare, even from the most minute dust particles, and the short focal length may result in poor illumination on objects that exceed a few microns in size or thickness. When choosing microscope slides for quantitative high-magnification darkfield microscopy, make certain to select slides made from a glass mixture that is free of fluorescent impurities.

High numerical aperture reflecting condensers (Figures 6.1, 6.7, and Table 6.2) with darkfield illumination provide the method of choice for observing and photographing collections of very small particles or colloidal suspensions, even when the particle diameter is significantly lower than the limit of resolution for the objective. This is due to light diffracted by the particles, which passes through the objective and becomes visible as bright diffraction disks. Each particle is visible as a minute diffraction disk, provided the lateral distance between adjacent particles is greater than the limit of resolving power of the objective. As illumination intensity is increased, the optical difference between minute diffracting particles and their background increases. Simultaneously, even smaller particles (detectable solely by their ability to scatter light) now diffract enough light to become visible and suspended particles can be seen even when their diameters are smaller than 40 nanometers, which is about one-fifth the 200 nanometer resolution limit with oil immersion objectives of the highest numerical

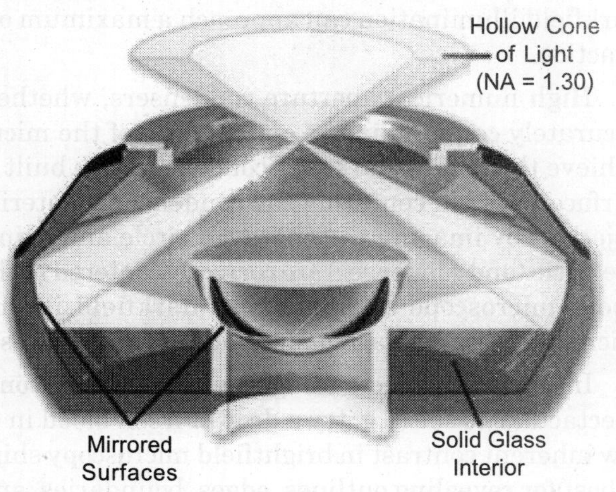

Fig. 6.7. Bispheric Double Reflecting Condenser.

aperture. In biological applications, the movements of living bacterial flagella that average about 20 nanometers in diameter (too small to be seen in brightfield or DIC illumination) can be observed and photographed using high numerical aperture darkfield condensers.

Careful attention should be paid to the details of oiling a high numerical aperture condenser to the bottom of the specimen slide. It is very difficult to avoid introduction of tiny air bubbles into the area between the condenser top lens and the bottom of the microscope slide, and this technique should be practiced to perfection. Air bubbles will cause image flare and distortion, leading to a loss of contrast and overall image degradation. Problems are also encountered when using microscope slides that are either too thick or too thin. Many darkfield condensers contain the range of usable slide thickness inscribed directly on the condenser mount. If the slide is too thick, it is often difficult to focus the condenser without resorting to a higher viscosity immersion oil. On the other hand, slides that are too thin have a tendency to break the oil bond between the condenser and the slide. It is a good idea to purchase precision microscope slides of the correct thickness to avoid any of the problems mentioned above.

A unique situation arises when specimens immersed in aqueous medium are being imaged using a high numerical aperture darkfield condenser. Under these conditions the refractive index of the aqueous solution limits the angle of inclination under which light can pass from the glass microscope slide (n = 1.515) into the water (n = 1.336) surrounding the specimen. The maximum numerical aperture of light passing from glass to water is given by the following equation:

$$\text{NA (illumination)} = 1.555 \times \sin(i) = 1.336 \times \sin(90°)$$

$$\text{and because } \sin(90°) = 1$$

$$\text{NA (illumination)} = 1.336$$

Even though reflecting darkfield condensers designed for oil immersion are listed with upper limits of numerical aperture as high as 1.50, light contributing to the illumination of specimens in aqueous media must have a numerical aperture no greater than 1.336, reducing the effective upper limit of darkfield illumination. In the case of specimens immersed in liquids of higher refractive index, the effective upper limit of the numerical aperture of darkfield illumination can approach a maximum of 1.50, although this is difficult to achieve in practice.

High numerical aperture condensers, whether intended for use dry or with oil, must be accurately centered in the optical path of the microscope to realize optimum performance. To achieve this, many darkfield condensers are built with a small circle engraved onto the upper surface to aid in centering the condenser. Centering is performed with a low power (10x-20x) objective by imaging the engraved circle and using the condenser centering screws to ensure the circle (and condenser) are correctly centered in the optical path. For more detailed information about microscope alignment for darkfield illumination, consult our section on darkfield microscope configuration elsewhere in the microscopy primer.

In general, objects imaged under proper conditions of darkfield illumination are quite spectacular to see (e.g. try a drop of fresh blood in darkfield). Often specimens containing very low inherent contrast in brightfield microscopy shine brilliantly in darkfield. Such illumination is best for revealing outlines, edges, boundaries, and refractive index gradients. Unfortunately, darkfield illumination is less useful in revealing internal details.

Chapter 7

The Fluorescent and Confocal Microscopy

INTRODUCTION TO FLUORESCENCE

The Physics of Light and Colour

When specimens, living or non-living, organic or inorganic, absorb and subsequently re-radiate light, the process is described as **photoluminescence**. If the emission of light persists for up to a few seconds after the excitation energy (light) is discontinued, the phenomenon is known as **phosphorescence**. Fluorescence, on the other hand, describes light emission that continues only during the absorption of the excitation light. The time interval between absorption of excitation light and emission of re-radiated light in fluorescence is of extraordinarily short duration, usually less than a millionth of a second.The phenomenon of fluorescence was known by the middle of the nineteenth century. British scientist Sir George G. Stokes first made the observation that the mineral **fluorspar** exhibits fluorescence when illuminated with ultraviolet light, and he coined the word "fluorescence". Stokes observed that the fluorescing light has longer wavelengths than the excitation light, a phenomenon that has become to be known as the **Stokes shift**. In Fig. 7.1, a photon of ultraviolet radiation (purple) collides with an electron in a simple atom, exciting and elevating the electron to a higher energy level. Subsequently, the excited electron relaxes to a lower level and emits light in the form of a lower-energy photon (red) in the visible light region. Fig. 7.2 is a diagrammatic representation of the visible light region of electromagnetic radiation, which covers a wavelength range from approximately 400 to 700 nanometers. Surrounding the visible region is higher

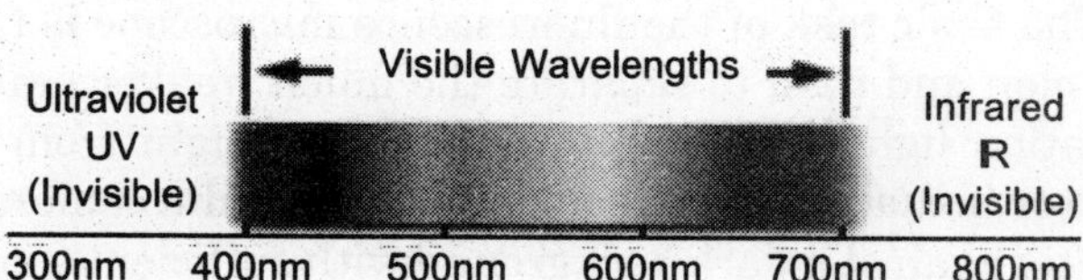

Fig. 7.1. Spectrum of "White" Light.

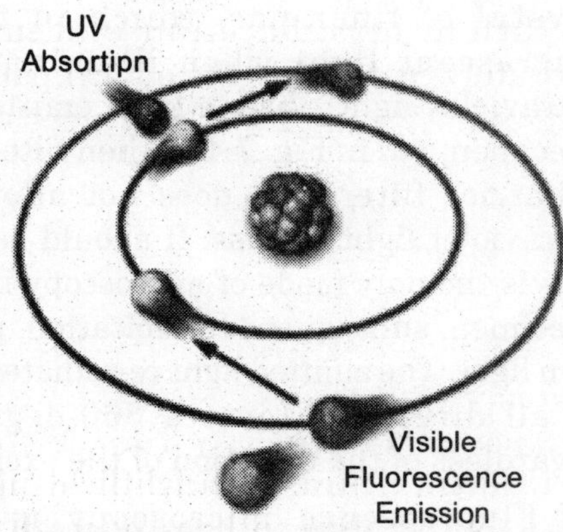

Fig. 7.2. Stokes' Observation.

energy ultraviolet light and lower energy infrared light.

Fluorescence microscopy is an excellent method of studying material that can be made to fluoresce, either in its natural form (termed **primary** or **autofluorescence**) or when treated with chemicals capable of fluorescing (known as **secondary fluorescence**). The fluorescence microscope was devised in the early part of the twentieth century by August Köhler, Carl Reichert, and Heinrich Lehmann, among others. However, the potential of this instrument was not realized for several decades, and fluorescence microscopy is now an important (and perhaps indispensable) tool in cellular biology.

Early investigations revealed that many specimens, including microminerals, crystals, resins, crude drugs, butter, chlorophyll, vitamins, and inorganic compounds, exhibit autofluorescence when irradiated with ultraviolet light. However, it was not until the 1930's that Austrian investigator Max Haitinger and other scientists developed the technique of secondary fluorescence, which employs fluorochrome stains to label specific tissue components, bacteria, and other pathogens that do not autofluoresce. These fluorochrome stains, tagged to specific biochemical targets, spurred the use of the fluorescence microscope. The instrument's value was significantly enhanced by the 1950's when Albert Coons and Nathan Kaplan demonstrated the localization of antigens in tissues that were reacted with a fluorescein-labeled antibody.

The basic task of the fluorescence microscope is to permit excitation light to irradiate the specimen and then to separate the much weaker emitted fluorescent light from the brighter excitation light. Thus, only the emission light from the specimen reaches the eye or other detector (usually a digital or conventional film camera). The resulting fluorescing areas shine brightly against a dark background with sufficient contrast to permit detection. The darker the background behind the non-fluorescing material, the more efficient the instrument.

Figure 7.3 presents a graphical representation of the occurrences when a fluorescent specimen is observed with a fluorescence microscope. Ultraviolet (**UV**) light of a specific wavelength or set of wavelengths is produced by passing light from an ultraviolet-emitting source through the exciter filter. The filtered ultraviolet light illuminates the specimen, in this case a crystal of fluorspar, which in turn emits fluorescent light when illuminated by the ultraviolet light. Visible light emitted from the specimen, red in Fig. 7.4, is then filtered through a barrier filter that does not allow reflected ultraviolet light to pass. It should be noted that this is the only mode of microscopy in which the specimen, subsequent to excitation, produces its own light. The emitted light re-radiates spherically in all directions (over a 360-degree angle), regardless of the direction of the exciting light.

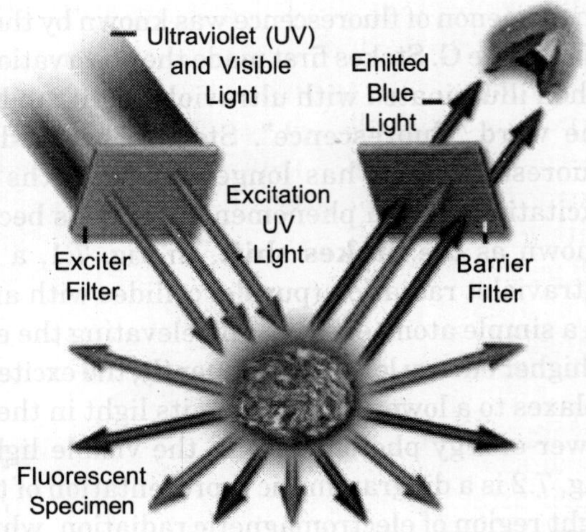

Fig. 7.3. Principle of Excitation and Emission.

Fluorescence microscopy is a rapidly expanding and invaluable tool of investigation. Its advantages are based upon attributes not as readily available in other optical microscopy techniques. The use of fluorochromes has made it possible to identify cells and sub-microscopic cellular components and other entities with a high degree of specificity amidst non-fluorescing material. What is more, the fluorescence microscope can reveal the presence of fluorescing material with exquisite sensitivity. An extremely small number of fluorescent molecules (as few as 50 molecules per cubic micrometer) can be detected. In a given sample, through the use of multiple staining, different probes will reveal the presence of individual target molecules. Although the fluorescence microscope cannot provide spatial resolution below the diffraction limit of the respective specimens, the presence of fluorescing molecules below such limits is made remarkably visible.

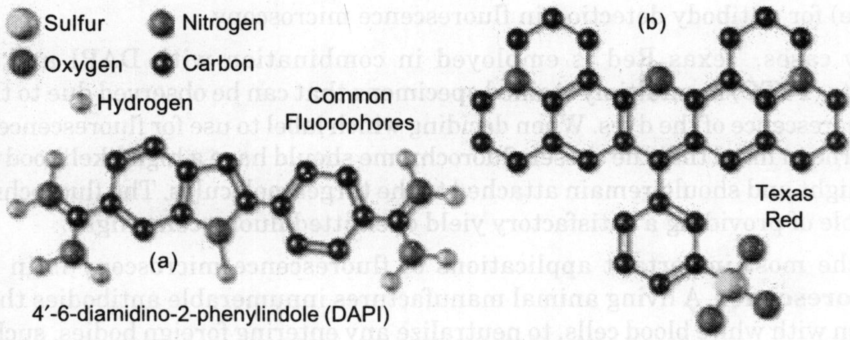

Fig. 7.4.

The techniques of fluorescence microscopy can be applied to organic material, formerly living (biological) material, or to living material (with the use of *in vitro* or *in vivo* fluorochromes) or to inorganic material (especially in the investigation of contaminants on semiconductor wafers). There are also a burgeoning number of studies using fluorescent probes to monitor rapidly changing physiological ion concentrations, such as calcium and magnesium, and pH values in living cells.

Many plant and animal tissues, as well as materials specimens, inherently fluoresce when irradiated with shorter wavelength light (primary or autofluorescence). Autofluorescence has been found useful in plant studies, coal petrography, sedimentary rock petrology, and in the semiconductor industry. In the study of animal tissues or pathogens, autofluorescence is often either extremely faint or of such non-specificity as to make autofluorescence of minimal utility. Of far greater value for such specimens are the extrinsic fluorochromes (also called fluorophores) that are excited by irradiating light, and whose eventual yield of emitted light is of greater intensity. Such fluorescence is known as secondary fluorescence.

Fluorochromes are stains, somewhat similar to the better-known visible light absorbing tissue stains, which attach themselves to visible or sub-visible organic matter. These fluorochromes, capable of absorbing and then re-radiating light, are often highly specific in their attachment targeting and have significant yield in absorption-emission ratios (a concept termed **quantum yield**). This makes fluorochromes extremely valuable in biological applications. The growth in the use of fluorescent microscopes is closely linked to the development of hundreds of fluorochromes with known intensity curves of excitation (absorption) and emission and well-understood biological structure targets.

There are two of the most commonly used fluorochromes in fluorescence microscopy. The popular nucleic acid stain 4',6-diamidino-2-phenylindole (**DAPI**), the left-hand molecule in Figure 4, is an indole derivative featuring two highly nucleophilic amidino moieties, which is useful for fluorescent labeling of nucleic acids and has been widely used in the fluorescence analysis for rapid identification of pathogens. Originally synthesized as a potential anti-trypanosomal agent, DAPI binds preferentially to adenosine and thymidine (**A-T**) base pair regions in DNA and is excited by ultraviolet light with a maximum absorption wavelength of 355 nanometers. The dye fluoresces in the blue region of the visible light spectrum. To the right of the DAPI molecule is another fluorochrome, Texas Red, whose properties in fluorescent conjugates have been thoroughly studied. This fluorochrome was developed for double labeling in an application known as **flow cytometry**, but is extensively used in place of rhodamine (another common fluorochrome) for antibody detection in fluorescence microscopy.

In many cases, Texas Red is employed in combination with DAPI and fluorescein isothiocyanate (**FITC**) for multiply stained specimens that can be observed due to the red, blue, and green fluorescence of the dyes. When deciding which label to use for fluorescence microscopy, it must be borne in mind that the chosen fluorochrome should have a high likelihood of absorbing the exciting light and should remain attached to the target molecules. The fluorochrome should also be capable of providing a satisfactory yield of emitted fluorescence light.

One of the most important applications of fluorescence microscopy is in the field of **immunofluorescence**. A living animal manufactures innumerable antibodies that are used, in conjunction with white blood cells, to neutralize any entering foreign bodies, such as viruses, bacteria, and foreign proteins, which contain or produce antigens. The antigen-antibody reaction is highly specific, often likened diagrammatically to a lock and key relationship. Immunofluorescence owes its success to the marriage between the sensitivity of fluorescence microscopy and the high degree of specificity exhibited by immunology.

In direct immunofluorescence, a specific antibody is labeled by chemically attaching a fluorochrome to form what is known as a **conjugate**, which is then spread on a microscope slide containing the suspected presence of a particular antigen known to stimulate production of the antibody. If the antigen is present, the labeled antibody conjugate binds to the antigen and remains bound to the specimen after it is washed. The presence of the chemically attached fluorescent conjugate and antigen is demonstrated when the fluorochrome is excited at its excitation peak, and the subsequent emission intensities at various wavelengths can then be observed visually or captured by a detector system (digital or traditional camera).

Another commonly used technique is known as **indirect** immunofluorescence, where serum possibly containing unlabeled antibody and its related, but known, antigen are incubated together. A fluorochrome conjugated to an anti-human antibody (if the specimen being tested is human) is then placed on the slide containing the unlabeled antibody-antigen. If indeed, there has been an antigen-antibody reaction, the fluorochrome-labeled anti-human antibody attaches itself to the complex formed by the antigen and antibody. Subsequently, the labeled grouping of antigen, antibody, and fluorochrome labeled anti-human antibody is excited at the peak wavelength intensity for that fluorochrome and any resulting emission is observed. The indirect immunofluorescence technique reduces the necessity of keeping in stock large numbers of labeled antibodies, and also usually results in greater fluorescence intensity.

A significant area of fluorescence investigation deals with cytochemical and histochemical staining. Fluorochromes have been used to identify chromosomes, DNA content, proteins, cellular structures, hormones, and vitamins. The fluorescence microscope is a powerful tool in such studies because of the exquisite sensitivity of selected fluorochromes and their specificity for extremely minute quantities in the specimen. Indeed, although the fluorescence microscope is limited in its spatial resolution to the usual rules governed by numerical aperture and diffraction limits, fluorescence probes can, through emitted light, reveal the presence of fluorescing material by making such material visible even when present in only sub-resolution quantities (such as a few molecules).

A burgeoning group of applications for fluorescence microscopy involve the application of fluorescent probes (fluorochromes) to living materials, both ***in vitro*** (in glass) and ***in vivo*** (in life). The difficulties multiply for these probes because of constraints imposed by possible toxicity. In addition, a considerable amount of attention must be paid to time intervals because of the ever-changing nature of the life processes and movement of intracellular structures. Fluorescence investigations have been applied to changes in ion concentration, bound and unbound, for important biological ions such as hydrogen (pH), calcium, and magnesium.

Among the best known are studies of intracellular calcium using the popular fluorescent probe **Fura-2**. For this dye, dual excitation at approximately 340 and 380 nanometers (using a light chopper and dual excitation filters or monochromators) can be monitored at a single emission wavelength, which is measured independently for each excitation band. A host computer controlling the microscope is used to calculate the ratio (in a process known as **ratio imaging** fluorescence) of bound to free intracellular calcium as revealed by the changes in fluorescence emission intensity. The advantage of the ratio method is that essentially all factors are kept constant except the dual near-ultraviolet excitation wavelengths, each of which produce emission in the green part of the visible light spectrum. A similar type of ratio imaging is conducted with a probe known as **Indo-1**. For this fluorochrome, also used for determination of intracellular bound and unbound calcium, a single excitation wavelength is employed, but emission is measured at each of two wavelengths to distinguish bound from unbound calcium.

Overview of Fluorescence Excitation and Emission Fundamentals

Because of their novel electronic configurations, fluorochromes have unique and characteristic spectra for absorption (usually similar to excitation) and emission. These absorption and emission spectra show relative **Intensity of fluorescence**, with the relative intensity classically plotted on the vertical axis versus wavelength on the horizontal axis. For a given fluorochrome, the manufacturers indicate the wavelength for the peak of the illumination excitation intensity and the wavelength for the peak of fluorescence emission intensity. It is important to understand the origin of the graphs and curves displaying the excitation and emission spectra for a given fluorochrome.

In order to determine the **emission** spectrum of a particular fluorochrome, the wavelength of maximum absorption (usually the same as the excitation maximum) is determined and the fluorochrome is excited at that wavelength. The absorption spectrum of a typical fluorochrome is illustrated in Fig. 7.5 where the relative intensity of absorption is plotted against the measured wavelength. A monochromator (a device that allows narrow bands of light wavelengths to pass) is then used to scan the fluorescence emission intensity over the entire series of

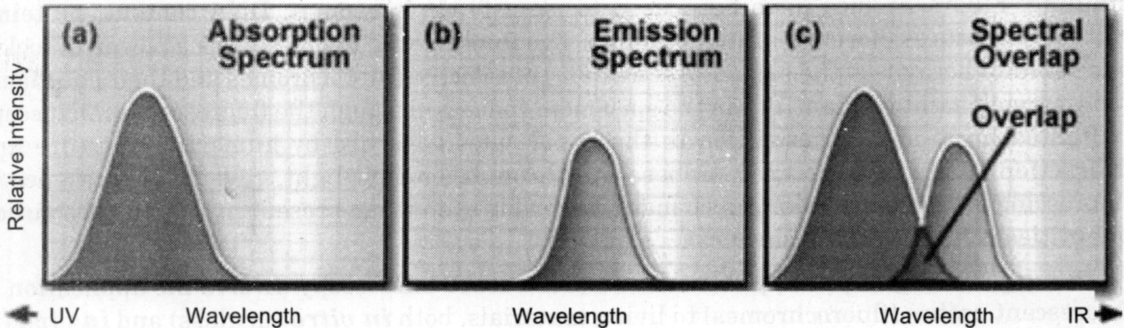

Fig. 7.5. The Absorption and the Emission Spectra with Overlap Profile.

emission wavelengths. The relative intensity of the fluorescence is measured at the various wavelengths to plot the emission spectrum, as illustrated in Fig. 7.5. The **excitation** spectrum of a given fluorochrome is determined in a similar manner by monitoring fluorescence emission at the wavelength of maximum intensity while the fluorophore is excited through a group consecutive wavelengths. The emission maximum is chosen and only emission light at that wavelength is allowed to pass to the detector. Excitation is induced (usually by means of a monochromator) at various excitation wavelengths and the intensity of the emitted fluorescence is measured as a function of wavelength. The result is a graph or curve, which depicts the relative **fluorescence** intensity produced by excitation over the spectrum of excitation wavelengths. Several observations can be made from a typical excitation and emission set of curves or spectra. There is usually an overlap between the higher wavelength end of the excitation spectrum and the lower wavelength end of the emission spectrum. This overlap of excitation and emission intensities and wavelengths (illustrated in Fig. 7.5 must be eliminated, in fluorescence microscopy, by means of the appropriate selection for an excitation filter, dichromatic beamsplitter (in reflected light fluorescence), and barrier or emission filter. Otherwise, the much brighter excitation light overwhelms the weaker emitted fluorescence light and significantly diminishes specimen contrast.When electrons go from the excited state to the ground state (see the section below entitled **Molecular Explanation**), there is a loss of vibrational energy. As a result, the emission spectrum is shifted to longer wavelengths than the excitation spectrum (wavelength varies inversely to radiation energy). This phenomenon is known as **Stokes Law** or **Stokes shift**. The greater the Stokes shift, the easier it is to separate excitation light from emission light. The emission intensity peak is usually lower than the excitation peak, and the emission curve is often a mirror image of the excitation curve, but shifted to longer wavelengths. In order to achieve maximum fluorescence intensity, the fluorochrome is usually excited at the wavelength at the peak of the excitation curve, and the emission detection is selected at the peak wavelength (or other wavelengths chosen by the observer) of the emission curve. The selections of excitation wavelengths and emission wavelengths are controlled by appropriate filters. In determining the spectral response of an optical system, technical corrections are required to take into account such factors as glass transmission and detector sensitivity variables for different wavelengths.

A typical fluorochrome absorption-emission spectral diagram is illustrated in Fig. 7.6. Note that the curves of fluorescence intensity for absorption (usually similar to the excitation curve for pure compounds) and emission for this typical fluorochrome are somewhat similar in shape. The wavelength shift between excitation and emission has been known since the middle of the nineteenth century (Stokes Law). Also note that the excitation and emission curves overlap somewhat at the upper end of the excitation and the lower wavelengths of the emission curve.

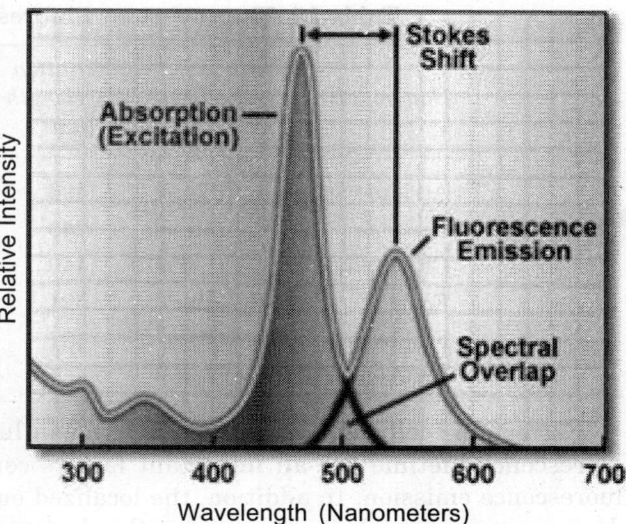

Fig. 7.6. Excitation and Emission Spectral Profiles.

The separation of excitation and emission wavelengths is achieved by the proper selection of filters to block or pass specific wavelengths of the spectrum as presented in Fig. 7.7. The design of fluorescence illuminators is based on control of excitation light and emission light by readily changeable filter insertions into the light path on the way toward the specimen and then emanating from the specimen. It is important, in view of low emission intensities, that the light source chosen for excitation be of sufficient brightness so that the relatively weak emission light can be maximized, and that fluorochromes of satisfactory absorption and yield be chosen.

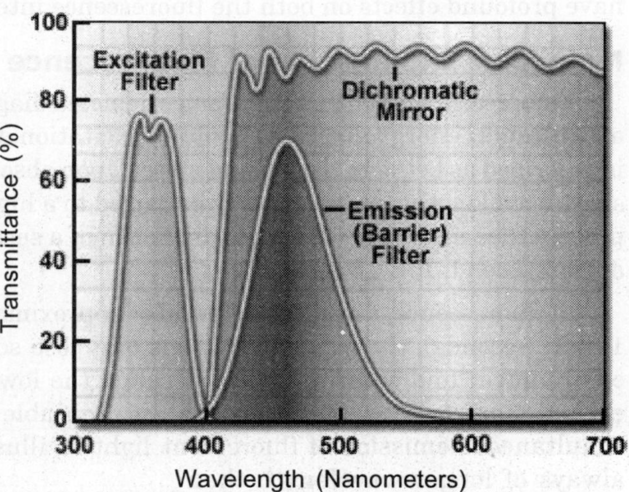

Fig. 7.7. Fluorescence Filter Spectral Profiles.

The efficiency with which the fluorochrome absorbs the excitation light is known as the extinction coefficient. The greater the extinction coefficient, the greater the possibility of light absorption in a given wavelength region (a prerequisite to ensuing fluorescence emission). The yield of emitted light is referred to as the quantum yield, the ratio of the number of quanta ("packets" of energy) emitted compared to the number of quanta absorbed (usually the yield is between 0.1 and 0.9). Quantum yields below 1 are the result of the loss of energy through non-radiative pathways (such as heat or a photochemical reaction) rather than the re-radiative pathway of fluorescence. Listed below in Table 7.1 are fluorescence quantum yields for a group of selected fluorochromes. Notice that some quantum yields seem trivial (benzene) while others are very efficient (Fluorescein and Rhodamine-B).

Table 7.1. Fluorochrome Fluorescence Quantum Yields.

Compound	*Solvent*	*Excitation Wavelength (nm)*	*Emission Wavelength (nm)*	*Quantum Yield*
Acridine Orange	Ethanol	493	535	0.46
Benzene	Ethanol	248	300-350	0.04
Chlorophyll-A	Ethanol	440	685	0.23
Eosin	Water	521	544	0.16
Fluorescein	Water	437	515	0.92
Rhodamine-B	Ethanol	555	627	0.97

Extinction coefficient, quantum yield, mean luminous intensity of the light source, and fluorescence lifetime are all important factors contributing to the intensity and utility of fluorescence emission. In addition, the localized environment surrounding the fluorochrome plays a paramount role in determining the characteristics of fluorescence emission. Variables such as solvent viscosity, ionic concentrations, pH, and hydrophobicity in the environment can have profound effects on both the fluorescence intensity and the lifetime of the excited state.

Molecular Explanation of Fluorescence

Fluorescence activity is sometimes depicted diagrammatically as shown in Fig. 7.8a (termed a **Jablonski** energy diagram). Prior to excitation, the electronic configuration of the molecule is described as being in the ground state. Upon absorbing a photon of excitation light, usually of short wavelengths, electrons may be raised to a higher energy and vibrational excited state, a process that may only take a quadrillionth of a second (a time period commonly referred to as a femtosecond, 10E-15 seconds).

In fluorescence, during an interval of approximately a trillionth of a second (a picosecond or 10E-12 seconds), the excited electrons may lose some vibrational energy to the surrounding environment and return to what is called the lowest excited singlet state. From the lowest excited singlet state, the electrons are then able to "relax" back to the ground state with simultaneous emission of fluorescent light as illustrated in Fig. 7.8(a). The emitted light is always of longer wavelength than the excitation light (Stokes Law) and continues so long as the excitation illumination bathes the fluorescent specimen. If the exciting radiation is halted, the fluorescence ceases.

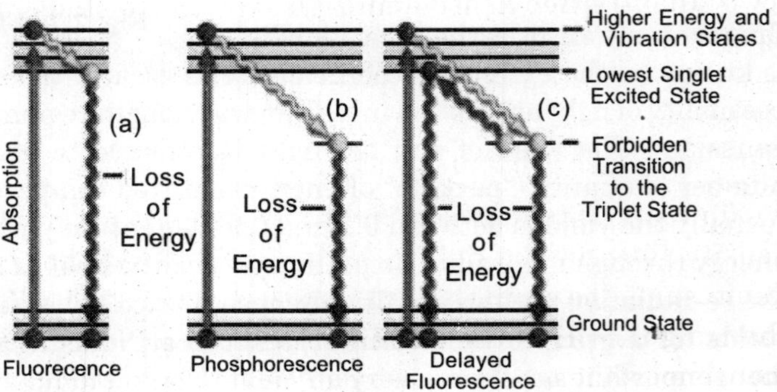

Fig. 7.8. Jablonski Energy Diagrams.

Occasionally the excited electrons, instead of relaxing to the lowest singlet state through vibrational interactions, make a **forbidden transition** to the exited triplet state and then to the ground state in a process where the emission of radiation may be considerably delayed-up to several seconds or more. This phenomenon is characteristic of phosphorescence, as shown in Figure 7.8(b). In some instances, the excited electrons may go from the triplet state back to the lowest excited singlet state and then return to the ground state, subsequently emitting fluorescent light. This action takes a little longer (about a microsecond or two) than usual fluorescence and is called delayed fluorescence (Fig. 7.8(c)). Under other circumstances (for example, photobleaching or the presence of salts of heavy metals or other chemicals), emitted light may be significantly reduced or halted altogether, as discussed below.

Fading or Photobleaching

There are specific conditions that may affect the re-radiation of light by an excited fluorophore, and thus reduce the intensity of fluorescence. This reduction of emission intensity is generally called **fading** or **photobleaching**. Some authors further subdivide fading into quenching and bleaching. Bleaching is irreversible decomposition of the fluorescent molecules because of light intensity in the presence of molecular oxygen. Quenching also results in reduced fluorescence intensity and frequently is brought about as a result of oxidizing agents or the presence of salts of heavy metals or halogen compounds.

Often, quenching results from the transfer of energy to other acceptor molecules physically close to the excited fluorophores, a phenomenon known as resonance energy transfer. This particular phenomenon has become the basis for a newer technique of measuring distances far below the lateral resolution of the light microscope.

The occurrence of bleaching has led to a technique known as **FRAP**, or fluorescence recovery after photobleaching. FRAP is based upon bleaching by short laser bursts and subsequent observation of the recovery of fluorescence caused by the diffusion of fluorophores into the bleached area.

Antifade ReagentCommentsp-phenylene-diamineThe most effective reagent for FITC. Also effective for Rhodamine. Should be adjusted to 0.1% p-phenylenediamine in glycerol/PBS for use. Reagent blackens when subjected to light exposure so it should be stored in a dark place. Skin contact is extremely dangerous.

DABCO(1,4-diazabi-cyclo-2,2,2-octane)Highly effective for FITC. Although its effect is slightly lower than p-phenylenediamine, it is more resistant to light and features a higher level of safety.n-propylgallateThe most effective reagent for Rhodamine, also effective for FITC. Should be adjusted to 1% propylgallate in glycerol/PBS for use.2-mercapto-ethylamineUsed to observe chromosome and DNA specimens stained with propidium iodide, acridine orange, or Chromomysin A3. Should be adjusted to 0.1mM 2-mercaptotheylamine in Tris-EDTA

To reduce the degree of fading in some specimens, it may be advisable to use a neutral density filter in the light path before the illumination reaches the excitation filter, thus diminishing the excitation light intensity. In other instances, fading effects may be reduced by changing the pH of the mounting medium or by using anti-bleaching agents (several of the more important agents are listed in Table 7.2). For digital imaging, photomicrography, or simply visual observation, rapidly changing the field of view may also avoid fading effects.

Table 7.2. Properties of Popular Antifade Reagents.

Antifade Reagent	*Comments*
p-phenylene-diamine	The most effective reagent for FITC. Also effective for Rhodamine. Should be adjusted to 0.1% p-phenylenediamine in glycerol/PBS for use. Reagent blackens when subjected to light exposure so it should be stored in a dark place. Skin contact is extremely dangerous.
DABCO (1,4-diazabi-cyclo-2,2,2-octane)	Highly effective for FITC. Although its effect is slightly lower than p-phenylenediamine, it is more resistant to light and features a higher level of safety.
n-propylgallate	The most effective reagent for Rhodamine, also effective for FITC. Should be adjusted to 1% propylgallate in glycerol/PBS for use.
2-mercapto-ethylamine	Used to observe chromosome and DNA specimens stained with propidium iodide, acridine orange, or Chromomysin A3. Should be adjusted to 0.1mM 2-mercaptotheylamine in Tris-EDTA

Practical Aspects of Fluorescence Filter Combinations

The terminology applied to fluorescence filters has become a jumble as a result of various initials utilized by different manufacturers to identify their filters. In this section, we attempt to make some order of this confusing terminology and specifically address how the various filter combinations can be utilized with specific fluorophores. Basically there are three categories of filters to be sorted out: **exciter** filters, **barrier** filters and **dichromatic** beamsplitters (dichroic mirrors), which are usually combined to produce a filter cube or block similar to the one illustrated in Fig. 7.9. Proper selection of filters is the key to successful fluorescence microscopy.

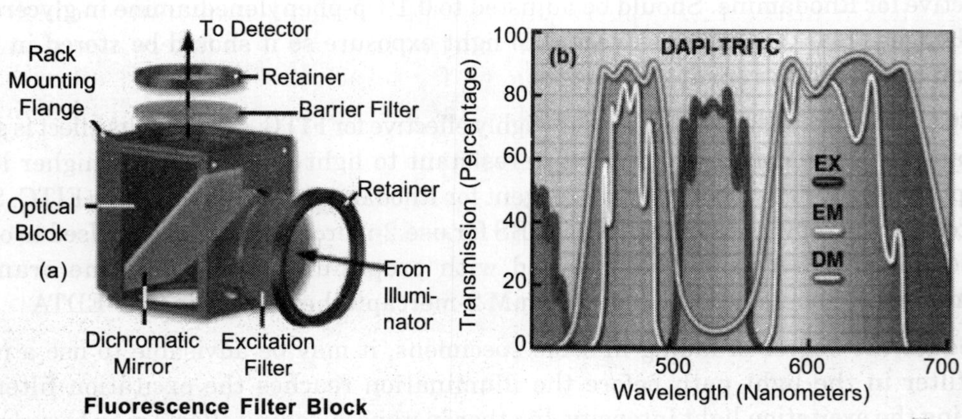

Fig. 7.9. Interference Filters for Fluorescence Microscopy.

The following sections contain links to review articles and interactive Java tutorials describing a variety of fluorescence filter combinations from the major manufacturers. Additional information is provided with regards to illumination excitation balancers, selection of fluorophores, and matching specific probes with filter combinations.

Brief Overview of Fluorescence Filters

Microscope manufacturers provide proprietary filter combinations (often referred to as **cubes** or **blocks**) that contain a combination of dichroic mirrors and filters capable of exciting fluorescent chromophores and diverting the resulting secondary fluorescence to the eyepieces or camera tube. A wide spectrum of filter cubes is available from most major manufacturers, which now produce filter sets capable of imaging most of the common chromophores in use today.

Basic Aspects of Light Filters

Many filters work by absorbing light, while others reflect unwanted light, but pass a selected region of wavelengths. The color temperature of light can be fine-tuned with filters to produce a spectrum of light having the characteristics of bright daylight, the evening sky, indoor tungsten illumination, or some variation in between. Filters are useful for adjusting the contrast of colored regions as they are represented in black and white photography or to add special effects in color photography. Specialized **dichroic** filters can be used to polarize light, while heat-absorbing filters can limit infrared wavelengths (and heat), allowing only visible light to pass through. Harmful ultraviolet rays can be exclusively removed from visible light by filters, or the intensity of all wavelengths (ultraviolet, visible, and infrared) can be reduced to specific ranges by neutral density filters. The most sophisticated filters operate by the principles of interference and can be adjusted to pass narrow bands (or even a single wavelength) of light while reflecting all others in a specific direction.

Interference Filters for Fluorescence Microscopy

The performance of high-resolution fluorescence microscopy imaging systems and related quantitative applications, especially as applied in living cell and tissue studies, requires precise optimization of fluorescence excitation and detection strategies. Fluorescence microscopy techniques could not have advanced so dramatically in recent years without significant developments in every dimension of the current state of the art, including the optical microscopes, the biology and chemistry of fluorophores, and perhaps most important, filter technology. The utilization of highly specialized and advanced thin film **interference filters** has enhanced the versatility and scope of fluorescence techniques, far beyond the capabilities afforded by the earlier use of gelatin and glass filters relying on the absorption properties of embedded dyes.

Acousto-Optic Tunable Filters (AOTFs)

Several benefits of the AOTF combine to greatly enhance the versatility of the latest generation of confocal instruments, and these devices are becoming increasing popular for control of excitation wavelength ranges and intensity. The primary characteristic that facilitates nearly every advantage of the AOTF is its capability to allow the microscopist control of the intensity and/or illumination wavelength on a pixel-by-pixel basis while maintaining a high scan rate. This single feature translates into a wide variety of useful analytical microscopy tools, which are even further enhanced in flexibility when laser illumination is employed.

Introduction to Fluorophores

Widefield fluorescence and laser scanning confocal microscopy rely heavily on secondary fluorescence emission as an imaging mode, primarily due to the high degree of sensitivity afforded by the techniques coupled with the ability to specifically target structural components and dynamic processes in chemically fixed as well as living cells and tissues. Many fluorescent probes are constructed around synthetic aromatic organic chemicals designed to bind with a biological macromolecule (for example, a protein or nucleic acid) or to localize within a specific structural region, such as the cytoskeleton, mitochondria, Golgi apparatus, endoplasmic reticulum, and nucleus. Other probes are employed to monitor dynamic processes and localized environmental variables, including concentrations of inorganic metallic ions, pH, reactive oxygen species, and membrane potential. Fluorescent dyes are also useful in monitoring cellular integrity (live versus dead and apoptosis), endocytosis, exocytosis, membrane fluidity, protein trafficking, signal transduction, and enzymatic activity. In addition, fluorescent probes have been widely applied to genetic mapping and chromosome analysis in the field of molecular genetics.

Spectral Bleed-Through Artifacts in Confocal Microscopy

The spectral bleed-through of fluorescence emission (often termed **crossover** or **crosstalk**), which occurs due to the very broad bandwidths and asymmetrical spectral profiles exhibited by many of the common fluorophores, is a fundamental problem that must be addressed in both widefield and laser scanning confocal fluorescence microscopy. The phenomenon is usually manifested by the emission of one fluorophore being detected in the photomultiplier channel or through the filter combination reserved for a second fluorophore. Bleed-through artifacts often complicate the interpretation of experimental results, particularly if subcellular colocalization of fluorophores is under investigation or quantitative measurements are necessary, such as in resonance energy transfer (**FRET**) and photobleaching (**FRAP**) studies.

Interference Filters

Recent technological achievements in bandpass filter design have led to the relatively inexpensive construction of thin-film interference filters featuring major improvements in wavelength selection and transmission performance. These filters operate by transmitting a selected wavelength region with high efficiency while rejecting, through reflection and destructive interference, all other wavelengths. Explore how interference filters operate by selectively transmitting constructively reinforced wavelengths while simultaneously eliminating unwanted light with this interactive tutorial.

Matching Fluorescence Filter Blocks with Fluorescent Probes

The essential feature of any fluorescence microscope is to provide a mechanism for excitation of the specimen with selectively filtered illumination followed by isolation of the much weaker fluorescence emission using a second filter to enable image formation on a dark background with maximum sensitivity. These conditions are satisfied in modern fluorescence instruments by a combination of filters that coordinate excitation and emission requirements based on the action and properties of the dichromatic beamsplitter. This interactive tutorial enables visitors to determine the optimum fluorophore characteristics necessary to maximize the efficiency of excitation and emission in conjunction with current Olympus fluorescence filter blocks.

Fluorescent Probe Excitation Efficiency

The absorption and fluorescence emission spectral profiles of a fluorophore are two of the most important criteria that must be scrutinized when selecting probes for applications in laser scanning confocal microscopy. In addition to the wavelength range of the absorption and emission bands, the molar extinction coefficient for absorption and the quantum yield for fluorescence emission should be considered. At laser excitation levels that do not saturate the fluorophore, fluorescence intensity is directly proportional to the product of the extinction coefficient and the quantum yield. This interactive tutorial examines how this relationship can be utilized to match fluorophores with specific lasers for confocal microscopy.

Acousto-Optic Tunable Filters

Wavelength selection is of fundamental importance in many arenas of the optical sciences, including fluorescence spectroscopy and microscopy. Electro-optic devices, such as the acousto-optic tunable filter (**AOTF**), are increasingly being employed to modulate the wavelength and amplitude of illuminating laser light in the latest generation of confocal microscopes. These filters do not suffer from the mechanical constraints, speed limitations, image shift, and vibration associated with rotating filter wheels, and can easily accommodate several laser systems tuned to different output wavelengths. In addition, acousto-optic filters do not deteriorate when exposed to heat and intense light as do fluorescence interference filters.

Liquid Crystal Tunable Filters

Liquid crystal tunable filters (**LCTFs**) use electrically controlled liquid crystal elements to select a specific visible wavelength of light for transmission through the filter at the exclusion of all others. This type of filter is ideal for use with electronic imaging devices, such as charge-coupled devices (**CCDs**), because it offers excellent imaging quality with a simple linear optical pathway.

Fluorochrome and Fluorescence Filter Combination Data Tables

In this section, we have compiled a listing of data on the most popular fluorochromes including excitation and emission characteristics, suggested cube utilization, and application data. This information can be accessed through the links provided below.

Fluorochrome Data Table

As a guide to fluorophores for confocal and widefield fluorescence microscopy, the table presented in this section lists many commonly-used fluorochromes, with their respective peak absorption and emission wavelengths and suggested laser illumination sources. While the authors assume responsibility for the accurate reporting of the data as published in various reliable sources, several caveats must be given. Fluorochrome dyes are environmentally sensitive and different results will be obtained with different solvents and applications. In reviewing the literature, one will frequently find somewhat different data supplied for the identical fluorochrome. There are also, in many instances, several sub-varieties of a fluorochrome.

- **Ex/Em/Cubes/Applications** - A comprehensive table of excitation and emission spectral maxima, and excitation suggestions, listed by application.
- **Excitation and Emission Data** - Excitation and emission maximum wavelengths listed alphabetically by fluorochrome name.
- **Olympus Fluorescence Vertical Illuminator Cubes** - Data tables of the Olympus cubes for older models, including the **U-URA, BH-2, Vanox-3**, and **IMT-3** fluorescent illuminators.

- **Fluorochromes and Suitable Cubes** - A compilation of spectral characteristics of major fluorochromes as well as sample fluorescence photomicrographs are also included in a short to long wavelength sequence.
- **Olympus UIS Fluorescence Mirror Units** - Designed to present the technical specifications of UIS fluorescence mirror units, this brochure offers a detailed unit selection chart based on fluorochrome characteristics. Mirror unit descriptions, a graphic representation of the characteristics of mirror unit wavelength, as well as information on custom fluorescence mirror units, fluorescence excitation balancers, and light source spectral characteristics are also provided.

Light Sources

In most fluorescence microscopy applications, the number of photons reaching the eye or other detector, such as a video camera or photomultiplier, is usually very low. This is because the quantum yield of most fluorochromes is low (quantum yield is the ratio of the number of quanta emitted by the specimen as compared to the number of quanta absorbed). To generate enough excitation light intensity to furnish emission capable of detection, powerful light sources are needed, usually arc (burner) lamps similar to the example that is presented in Fig. 7.10. These lamps are filled with high pressure gases and should be handled with caution.

Fig. 7.10. Mercury arc lamp.

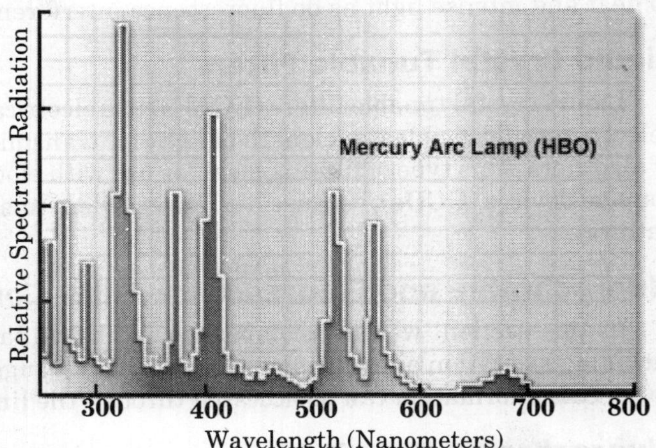

Fig. 7.11. Mercury arc lamp UV and visible emission spectrum.

The most common lamps are the mercury burners, ranging in wattage from 50 watts to 200 watts and the xenon burners ranging from 75 watts to 150 watts. The mercury burner lamp consists of two electrodes sealed under high pressure in a quartz glass envelope which also contains mercury.

These light sources are usually powered by a D.C. power supply furnishing enough start-up power to ignite the burner (by ionization of the gaseous vapor) and to keep it burning with a minimum of flicker. The power supply should have a timer to enable you to keep track of the number of hours the burner has been in use. Arc lamps lose efficiency and are more likely to shatter, if used beyond their rated lifetime. The mercury burners do not provide even intensity across the spectrum from ultraviolet to infrared (See Fig. 7.12 for the emission spectrum of the mercury burner). Much of the intensity of the mercury burner is expended in the near ultraviolet, with peaks of intensity at 313, 334, 365, 406, 435, 546, and 578 nanometers. At other wavelengths of visible light, the intensity is steady but not nearly so bright, but still usable for blue excitation. It also should be understood that mere wattage is not the only consideration for determining brightness.

Fluorescence Lamp Specifications (Table)

Another important criterion is the size of the arc; the brightness per unit area of the arc encompassed within the back aperture of the objective is a better measure of the useful brightness of the burner. Using the criterion, you will note that of the three mercury burners listed in Table 7.3, the 100 watt burner (other things being equal) is the brightest of the three. Also important in your selection of a burner is whether or not the spectral intensity peaks of the burner's wavelengths match the excitation requirements for the fluorochromes staining the specimen. Whenever an illumination system is being evaluated, it is necessary to consider the entire system including collector lenses and the use of aperture and field diaphragms in securing Köhler illumination.

Table 7.2. Fluorescence Lamp Specifications.

Lamp Type	*XBO 150W/1*	*XBO 75W/2*	*HBO 200W/2*	*HBO 100W/2*	*HBO 50W/3*
Current	DC	DC	DC	DC	DC
Rate Power (watts)	150	75	200	100	50
Luminous Flux (lumens)	3000	950	10000	2200	1300
Light Intensity (Candella)	300	100	1000	260	150
Avg. Brightness (cd/cm)	15000	40000	40000	170000	90000
Arc Size (w x h in millimeters)	0.50 x 2.20	0.25 x 0.50	0.60 x 2.20	0.25 x 0.25	0.20 x 1.35
Life (Hours)	1200	400	400	200	200

The xenon burners have much more even intensity across the visible spectrum than do the mercury burners (See Fig. 7.12); they do not have the very high spectral intensity peaks that are characteristic of mercury burners. Xenon lamps are deficient in the ultraviolet; they expend a large proportion of their intensity in the intra-red, and therefore the use of such lamps requires care in control of heat. Short-gap xenon burners are usually more desirable because the size of the arc is such that its light may be more readily included within the back aperture of the objective, thus avoiding waste of light intensity.

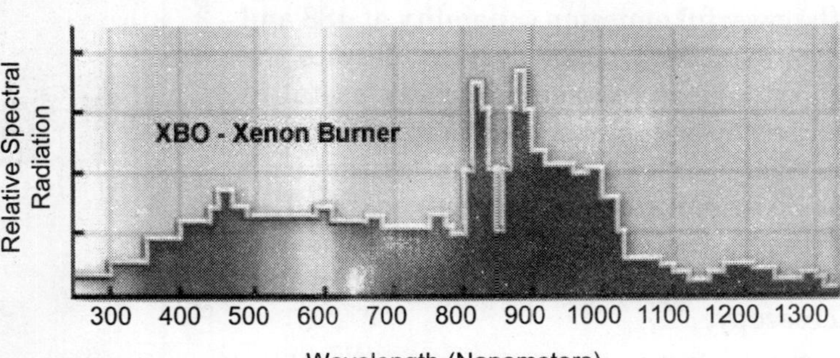

Fig. 7.12. Xdnon Arc Lamp Emission Spectrum.

Caution: *Mercury and Xenon arc lamps require caution during operation because of the danger of explosion due to very high internal gas pressures and extreme heat generated during use. Never ignite a lamp outside of its housing or observe the lamp directly when it is burning (this can cause serious eye damage). Neither mercury nor xenon lamps should be handled with bare fingers in order to avoid inadvertent etching of the quartz envelope. Change bulbs only after the lamp has had sufficient time to cool. Store lamps in their shipping containers to avoid accidents.*

Always adhere to the safety procedures listed above when installing or changing mercury or xenon arc lamps.

Mercury burners have a life of 200 hours; xenon burners several hundreds of hours. Frequent on-off switching reduces lamp life. When the burners reach their rated lifetime, the spectral emissions may change and the quartz envelope weakens.

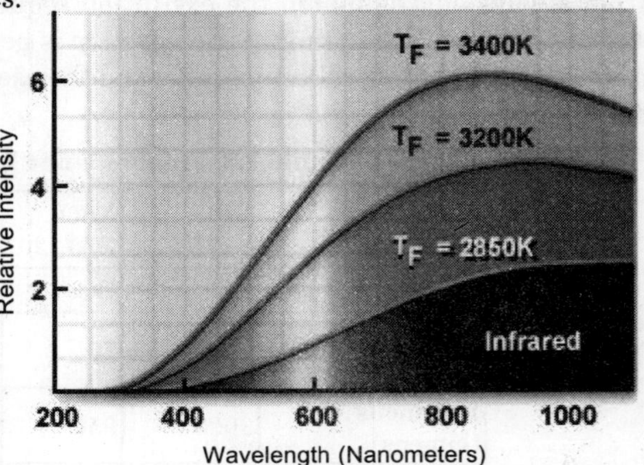

Fig. 7.13. Tungsten Lamp Emission Spectrum.

Sometimes, tungsten-halogen bulbs are used, especially for blue or green excitation with brightly emitting specimens. Their output is relatively even across the visible spectrum (Fig. 7.13); they are deficient in the near-ultraviolet and also have a relatively high proportion of intensity in the infrared. These lamps do not require expensive power supplies for ignition but are powered by low voltage transformers; the bulbs last for 30 to 50 hours when used at their maximum rated voltage.

In recent years, there has been increasing use of **lasers**, particularly the argon-ion laser with powerful emission capability at 488 and 514 nanometers. Laser sources, despite the high cost, have become especially useful in laser scanning confocal microscopy. There are a number of different types of lasers that each provide a unique emission spectrum. Fig. 7.14 illustrates the emission spectra of the two most commonly used lasers in fluorescence microscopy.

This technique, with its many variations of equipment, has proved to be a powerful tool in rendering very sharp fluorescence images by ingeniously controlling out-of-focus light. This is accomplished through scanning and imaging extremely small, shallow areas successively. The optical sections of the specimen are stored in a computer and

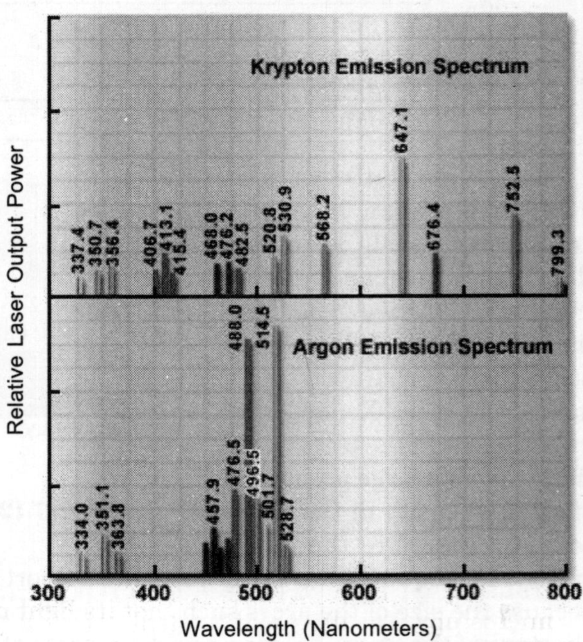

Fig. 7.14. Laser Illumination Source Emission Spectra.

reconstituted into the whole image which can then be displayed on a video monitor or printed with a video printer.

Lamp Alignment

Microscope companies may offer an optional centering screen to facilitate the centering of the image of the lamp arc to the back aperture of the objective. This accessory has, at its upper end, the standard Royal Microscopical Society (RMS) thread and can be screwed into the nosepiece. It is placed there and rotated into the light path. The lower face of the accessory has a frosted, orange-colored glass with an inscribed crosshair.

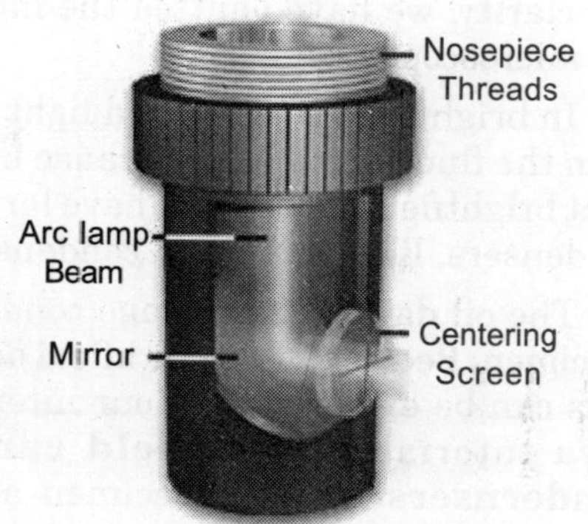

Fig. 7. 15. Arc Lamp Alignment Accessory.

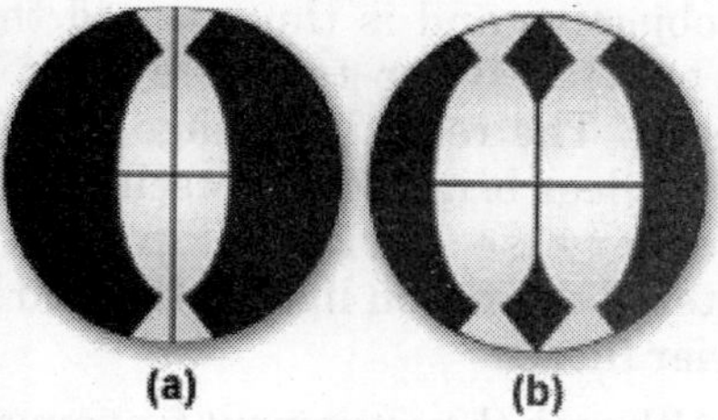

Fig. 7. 16.

Light coming down from the dichroic mirror strikes the built-in reflector of the centering screen and is reflected onto the crosshairs, as illustrated in Fig. 7.15. As you manipulate the lamp condenser knob and the centering screws on the lamp socket, you can move the image so that it is centered to the crosshairs (Fig. 7.16(a)). The size of the image of the arc can be made bigger or smaller by manipulating the lamp condenser lever. When you are done with the centering accessory, it can be replaced by a regular objective. If the lamphouse contains a mirror, the mirror position should be adjusted so that the arc image in the mirror is positioned parallel and adjacent to the arc image itself as shown in Fig. 7.15(b).

If you do not have a centering screen, the following alternative procedure can be used. Focus the specimen with the 10x objective. Then rotate the nosepiece so that an empty opening on the nosepiece is in the optical path of the microscope. Place a white card on the stage (in place of the specimen) and you will see the image of the arc projected onto the card. By manipulating the lamp condenser knob and the burner centering screws on the lamphouse, you can center the image of the arc to the optical axis of the microscope.

Transmitted Light Illumination

The early fluorescence microscope utilized transmitted light illumination (diascopic fluorescence). A primary filter to select the excitation light wavelengths was placed in the light port of the microscope and a secondary barrier filter was positioned above the microscope nosepiece to block residual excitation light and to select emission wavelengths reaching the eye or camera.

The optical pathways illustrated in Fig. 7.17 represent the two most common modes of transmitted light fluorescence. Brightfield fluorescence (Fig. 7.17(a)) is similar to standard brightfield microscopy with the addition of a barrier filter after the objective and a dichroic excitation filter before the condenser. To explore the pathway of visible light through a transmitted light microscope, visit our **interactive Java tutorial** on transmitted light microscopy. The darkfield condenser in Fig. 7.17(b) is a very efficient Tiyoda double lens and mirror condenser that projects light onto the sample at oblique angles, preventing excitation wavelengths from directly entering the objective. In this example, the condenser is fitted with an oil front lens and there is immersion oil between the microscope slide and the condenser (for clarity, we have omitted the immersion oil that should reside between the objective and the microscope slide).

In brightfield transmitted light fluorescence, it is difficult to separate the excitation light from the fluorescing light because both kinds of light directly enter the objective. Transmitted light brightfield condensers have largely been replaced by high numerical aperture oil darkfield condensers, like the Tiyoda condenser illustrated in Fig. 7.17(b).

The oil darkfield substage condenser directed excitation light at steep angles toward the specimen. Because of the darkfield design, most of the excitation light never enters the objective. This can be explored with our interactive Java tutorial on **darkfield cardioid condernsers**. As the specimen absorbs excitation light and emits only longer wavelength light, it is the longer wavelength light that gains admittance to the objective and is thus passed through the barrier filter to the eye or other detector. The resulting image shows as a more or less brightly fluorescing object on an otherwise dark background. Any scattered excitation light is blocked by the barrier filter.

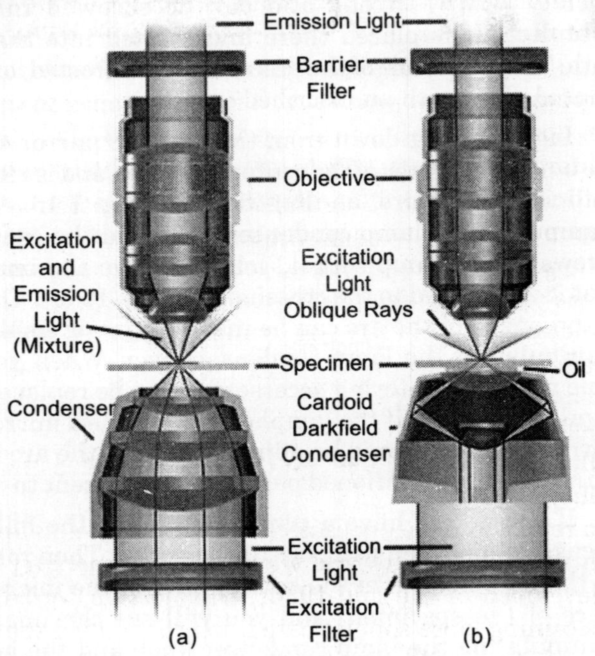

Fig. 7.17.

Although the equipment for transmitted light darkfield fluorescence is relatively simple, the technique has significant disadvantages. Many users find it difficult to properly align the oiled condenser to the optical axis of the microscope. In addition, the numerical aperture of the higher magnification oil or water immersion objectives has to be reduced by a built-in iris diaphragm (with consequent loss of light intensity and resolution) in order to prevent excitation light from entering the objective directly.

Transmitted light darkfield technique also precludes the use of simultaneous fluorescence viewing along with phase microscopy or Nomarski differential interference contrast microscopy. The darkfield method is also very wasteful of light, since the excitation light irradiates much of the specimen outside of the field of view being observed, thus reducing the usability of excitation intensity.

Anatomy of the Fluorescence Microscope

In contrast to other modes of optical microscopy that are based on macroscopic specimen features, such as phase gradients, light absorption, and birefringence, fluorescence microscopy is capable of imaging the distribution of a single molecular species based solely on the properties of fluorescence emission. Thus, using fluorescence microscopy, the precise location of intracellular components labelled with specific fluorophores can be monitored, as well as their associated diffusion coefficients, transport characteristics, and interactions with other biomolecules. In addition, the dramatic response in fluorescence to localized environmental variables enables the investigation of pH, viscosity, refractive index, ionic concentrations, membrane potential, and solvent polarity in living cells and tissues.

The earliest fluorescence microscope configurations featured a classical brightfield or darkfield diascopic (transmitted light) optical train that focused excitation light passed through a filter onto the specimen plane. Fluorescence emission was gathered by the objective, along with a significant amount of the excitation illumination, and projected through a second filter into the eyepiece diaphragm to form the intermediate image. Because the intensity of excitation light is usually several orders of magnitude greater than fluorescence emission, the specimen view in these early transmitted light microscopes was often very low in contrast and superimposed on a background flooded with scattered excitation illumination. Using a high numerical aperture oil-immersion darkfield condenser to illuminate the specimen at highly oblique azimuths helped to eliminate most of the background noise, but did not provide adequate illumination for any but the lowest numerical aperture objectives. The resulting images suffered from poor resolution and very low brightness levels.Fluorescence microscopy with **incident** (reflected light or **episcopic**) illumination was first developed in the late 1920s to observe fluorescence emission in opaque metallurgical specimens. Like their brightfield counterparts, these reflected light fluorescence instruments employed half-mirror beamsplitters that are restricted to an overall efficiency of approximately 25 per cent (after losing 50 per cent of the illumination intensity upon each pass through the mirror). However, reflected light fluorescence microscopes enjoyed the advantage of having a high numerical aperture objective acting as the condenser, and were able to produce images with a significantly greater level of brightness than transmitted light microscopes operating at similar numerical apertures. In addition, the stray excitation light reflecting back into the objective (at high numerical aperture) was reduced to only a few per cent. The dual role of the objective in reflected light microscopy also significantly eases the task of alignment. Merely focusing the objective onto the specimen establishes both the illumination field and the field of view such that incident excitation light and the observed fluorescent emission follow identical pathways through the microscope optical train.The move to reflected light fluorescence (at least for metallography) was bolstered by technical advances in illumination sources when compact mercury-vapor and xenon arc-discharge lamps were developed in the mid-1930s. During this period, colored glass and gelatin filters were also becoming more sophisticated, which enabled the application of fluorochromes excited by blue and green visible light using tungsten-halogen lamps. Anti-reflective coatings and improved glass formulations led to major improvements in objective design during the 1940s, but the most fundamental contribution to the development of incident light fluorescence microscopy was the introduction of **dichromatic** beamsplitters (also termed **dichroic mirrors**) by Russian optical scientist Eugeny Brumberg in 1948. This innovation overcame the light loss problems inherent with the use of ordinary half-mirrors in reflected light microscopy. Reflected light fluorescence microscopes were first commercialized on a broad scale by Johan S. Ploem in the late 1960s, who was instrumental in developing the Wild-Leitz **Ploem Opak**, containing multiple optical blocks that were interchangeable and housed various combinations of filters for fluorescence microscopy.

Basic Strategy of Epi-Fluorescence Microscopy

Reflected light fluorescence microscopy is overwhelmingly the current method of choice for widefield investigations with non-coherent light sources, as well as those conducted with laser scanning confocal and multiphoton instruments. This popular mode of fluorescence microscopy is also known as incident light fluorescence, episcopic fluorescence, or simply **epi-**

fluorescence. A typical modern reflected light fluorescence microscope, also equipped for transmitted light observation in a variety of contrast-enhancing modes, is presented in Fig. 7.18. The microscope contains a trinocular observation head that is coupled to a charge-coupled device (**CCD**) camera system, and has two illumination sources, one for transmitted light and the other for episcopic observations (tungsten-halogen and mercury arc-discharge, respectively). A microscope of this design can combine or alternate reflected light fluorescence with transmitted light phase contrast, differential interference contrast (**DIC**), polarized light, or Hoffman modulation contrast observation.

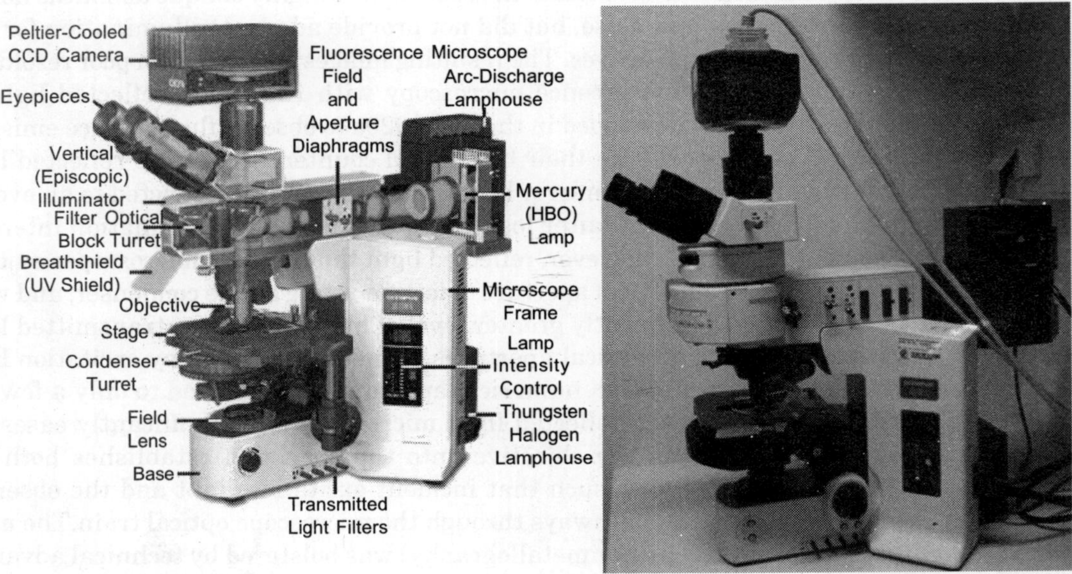

(a) The Internal System of A Fluorescent Microscope.

(b) The Actual Fluorescence Microscope.

Fig. 7.18.

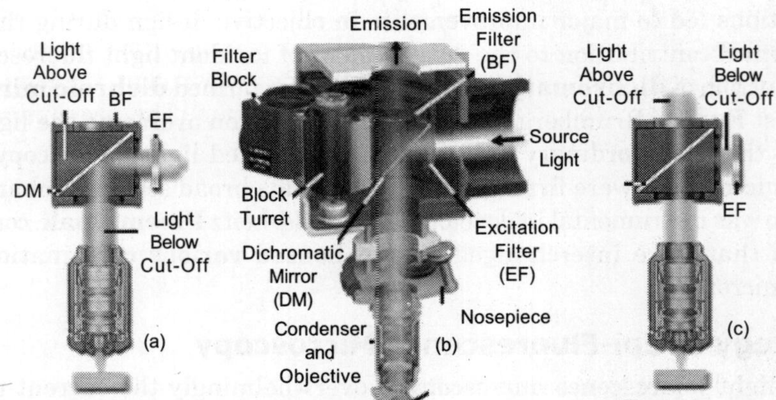

Fig. 7.19. Dichromatic Mirror Function in Reflected Light Fluorescence Illumination.

The essential feature of any fluorescence microscope is to provide a mechanism for excitation of the specimen with selectively filtered illumination followed by isolation of the much weaker fluorescence emission using a second filter to enable image formation on a dark background with maximum sensitivity. Localized probe concentration in biological specimens is so low in many experiments that only a small fraction of the excitation light is absorbed by the fluorescent species. Furthermore, of those fluorophores that are able to absorb a quantity of illumination, the per centage that will emit secondary fluorescence is even lower. The resulting fluorescence emission brightness level will range between three and six orders of magnitude less than that of the illumination. Thus, the fundamental problem in fluorescence microscopy is to produce high-efficiency illumination of the specimen, while simultaneously capturing weak fluorescence emission that is effectively separated from the much more intense illumination band. These conditions are satisfied in modern fluorescence instruments by a combination of filters that coordinate excitation and emission requirements based on the action and properties of the dichromatic beamsplitter. The principles behind dichromatic beamsplitter (mirror) function in reflected light fluorescence microscopy are outlined in Fig. 7.19 for a hypothetical specimen containing a fluorophore that is excited in the green region (550 nanometers) and fluoresces in the red (620 to 660 nanometers) wavelengths of the visible light spectrum. A high-intensity light source outputs a wide spectrum of excitation wavelengths at high flux density (usually covering most of the ultraviolet and the entire visible spectrum), which travels through the illuminator and first encounters a filter that selects the proper wavelength band for excitation (labeled **EF**; see Fig. 7.19(a)). In this case, the filter passes light having wavelengths between 510 and 560 nanometers with high efficiency, but also allows other wavelengths to pass at a much lesser extent. The excitation light next reaches the dichromatic mirror (**DM** in Figure 68) and is reflected into the objective rear aperture to form a cone of illumination that bathes the specimen. The dichromatic mirror is positioned in the light path at a 45-degree angle and designed to selectively reflect wavelengths between 490 and 565 nanometers (as illustrated in Fig. 7.20), while simultaneously transmitting both shorter and longer wavelengths.Presented in Figure 7.20 are the transmission profiles for the filter combination used to separate excitation illumination from fluorescence emission in Fig. 7.19. The excitation filter spectrum (red curve) exhibits a high level of transmission (approximately 80 per cent) between 510 and 560 nanometers with a center wavelength (**CWL**) of 535 nanometers. The dichromatic mirror (yellow curve) reflects wavelengths in the region of the excitation spectrum, while passing higher and lower wavelengths with relatively high efficiency. Note that zero per cent transmission on the dichromatic mirror curve corresponds to 100 per cent reflection. The pronounced dip in the transmission profile between 490 and 570 nanometers, which represents a peak in reflectance, serves to reflect the band of wavelengths passing from the excitation filter at a 90-degree angle and onto the specimen. The final component in the optical train, an emission or barrier filter (white curve), transmits wavelengths higher than 590 nanometers, which corresponds to visible light having yellow, orange, and red colors. Boundaries between transmitted and reflected wavelength bands of the various superimposed spectra are designed to be as steep as possible to assure nearly complete separation of the reflected and transmitted wavelengths. A pattern of sinusoidally rising and falling spikes appearing in the dichromatic mirror spectrum is a common effect of the thin-film deposition process known as **ringing**. The performance of this filter combination is remarkable and is a clear demonstration of the rapid advances being achieved in thin film interference filter technology.

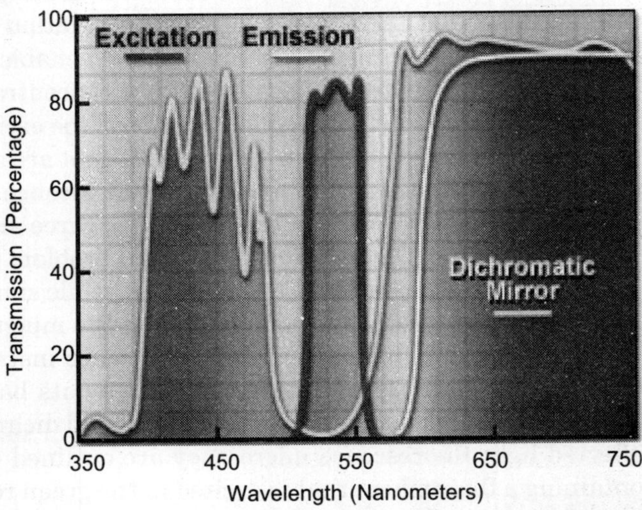

Fig. 7.20. Green Excitation and Red Emission Filter Set.

Because only a narrow bandwidth of light is reflected by the dichromatic mirror, illumination wavelengths shorter than 490 nanometers and longer than 565 nanometers that manage to pass through the excitation filter are also transmitted through the dichromatic mirror, illustrated as the light above the cut-off in Figure 7.19(a). Note that the reflection of excitation light is not 100 per cent efficient, and thus, a small amount of green light passes through the dichromatic mirror without being reflected. In addition, not all of the light having wavelengths above 565 or below 490 nanometers is transmitted through the mirror. A small per centage of this light is reflected by the mirror through the objective and onto the specimen. Light transmitted from the excitation filter through the dichromatic mirror is partially absorbed by the flat black coating on the interior of the filter block, but some reflects from the surface and passes through the barrier filter at an oblique angle, contributing to the fluorescence background noise. Fluorescence emission by the specimen (primarily red wavelengths), which results from the green light excitation, is gathered by the objective and passes through the dichromatic mirror and barrier filter (light above cut-off in Fig. 7.19(c)). The barrier filter (labelled **BF** in Fig. 7.19) is specifically designed to allow only light of wavelengths greater than 590 nanometers to reach the microscope eyepieces and/or detector. In serving this duty, the barrier filter effectively prevents excitation light wavelengths reflecting from the specimen (and successfully traversing the dichromatic mirror) from reaching the detector. However, a majority of the excitation wavelengths returning from the specimen are reflected towards the excitation filter and illuminator by the dichromatic mirror (the light below cut-off in Fig. 7.19(c)). The net effect of the filter configuration illustrated in Figures 68 and 69 is to separate the excitation light, which is substantially higher in intensity, from the much weaker fluorescence emission. In all cases, as shown in the figures, the cut-off levels of the filters involved in fluorescence microscopy are not absolute, but enable some light outside the wavelength range to bleed through. The entire sequence of events is graphically presented in Fig. 7.19(b), which illustrates the optical pathways and strategic filter arrangement necessary for efficient reflected light fluorescence microscopy.

The Vertical Fluorescence Illuminator

At the heart of the modern fluorescence microscope is the universal reflected light vertical illuminator, which is interposed between the observation viewing tubes and the nosepiece carrying the objectives, as illustrated in Figs. 7.18 and 7.21. The illuminator is designed to direct light generated by a high-intensity source onto the specimen by first passing the light through the microscope objective on the way toward the specimen and then using that same objective to capture the light being emitted by the specimen. This type of illumination strategy has several advantages. The microscope objective, which acts first by serving as a well-corrected condenser, next gathers image-forming fluorescence emission for transmission to the eyepieces or camera detection system. As such, the objective is always in correct alignment relative to each of these functions. Furthermore, most of the excitation light that is scattered or reflected by the specimen (over a 360-degree angle) travels away from the objective front lens element, rather than being projected directly into the glass, as is the case in transmitted fluorescence illumination. This effect is referred to as **front-face** illumination and is particularly useful with thick specimens. Finally, the specimen area being illuminated is restricted to the same area that is being observed, and both illumination and light collection can utilize the full numerical aperture of the objective.At the far end of the vertical illuminator is the lamphouse (see Fig. 7.21), which contains a high-intensity arc-discharge or filament-based incandescent light source. The most popular illumination source is a high-pressure 100-watt mercury arc lamp (**HBO**, often termed a **burner**), but xenon and metal halide arc lamps, lasers, and tungsten-halide incandescent lamps can serve the purpose as well. Light emitted by the source is focused by the collector lens system and travels along the interior of the illuminator parallel to the tabletop and perpendicular to the microscope optical axis. The illuminator design illustrated in Fig. 7.21 contains a multiple-component collector lens system, but a spherical lens elements are also being implemented to improve correction for chromatic aberration in the near-infrared and ultraviolet regions. A heat filter designed to remove or suppress infrared wavelengths is placed either inside the lamphouse itself or adjacent to the lamphouse mount at the rear of the vertical illuminator. In addition, some microscopes contain a dual or multi-band **excitation balancer** system (see Figures 7.21 and 7.23) near the lamphouse to selectively filter wavelengths that will not be needed for excitation in order to fine-tune the performance of the fluorescence filter set contained at the other end of the illuminator.

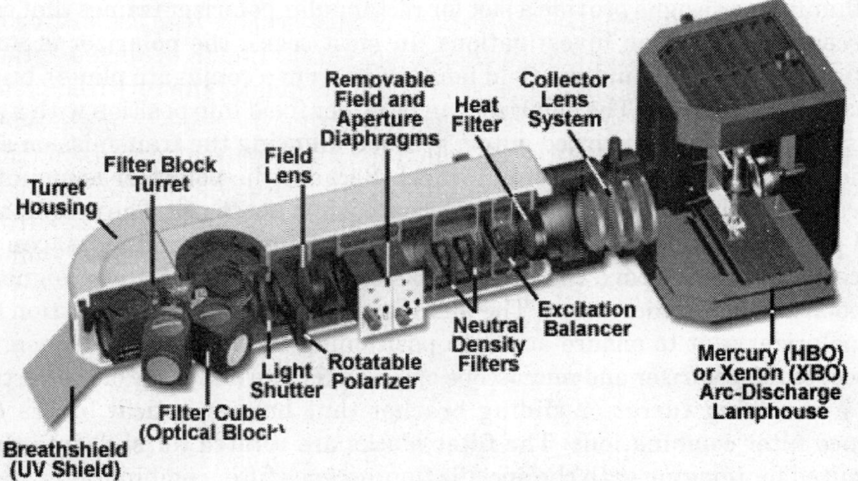

Fig. 7.21. Fluorescence Vertical (Episcopic) Illuminator.

Also positioned near the lamphouse in the vertical illuminator are a set of neutral density filters that can be employed to adjust the overall intensity of light passing through the system and reduce fluorescence fading or photobleaching. These filters are usually mounted on slider frames that enable them to be quickly inserted and removed from the light path while viewing the specimen fluorescence. In order to establish Köhler illumination, the vertical illuminator contains a centerable aperture and field diaphragm combination, both of which have variable iris apertures that dictate the field size and illumination intensity. Centering of the aperture and field diaphragms and adjustment of the iris size is accomplished with several pairs of knobs located on each exterior side of the vertical illuminator housing. In some designs, a slider knob is provided that removes the entire diaphragm assembly from the light path to maximize the amount of light passing through the illuminator. Other microscopes contain a pinhole aperture of fixed size mounted on a slider that can be inserted into the light path to dramatically restrict the illumination field for specialized applications, such as fluorescence recovery after photobleaching (**FRAP**) investigations.Following the field and aperture diaphragms in the vertical illuminator optical train is the field lens, which is necessary to spread the light and create a sufficient illumination field for establishing Köhler illumination. After the field lens, all modern vertical illuminators contain a light shutter (with manual on-off control) that blocks the intense excitation light from reaching the filter sets and specimen when fluorescence is not being observed or recorded. Shutters are a critical feature in fluorescence microscopy because continuous illumination of the specimen can dramatically reduce emission due to photobleaching, and high-intensity light is unhealthy for living cells. In addition, a shutter enables the arc-discharge lamp to remain active (alleviating a warm-up period) while specimens are being transiently examined with transmitted light. High-end fluorescence microscopes are often equipped with electronic shutters that provide rapid and remote control of illumination. An ultraviolet protection shield (labeled **breathshield** in Figure 7.21) is fitted onto the front of the illuminator housing to protect the operator from any inadvertent leakage of potentially dangerous short-wavelength ultraviolet radiation when the shutter is open. The shield serves double duty by also protecting the specimen from exhaled breath while under observation.Several vertical illuminator designs provide a slot for rectangular polarizer frames that can be employed in fluorescence polarization investigations. In most cases, the polarizer is inserted into the illumination light path behind the field lens (away from a conjugate plane), but preceding the light shutter (see Fig. 7.21). The polarizer can be either fixed into position with a pre-determined transmission azimuth, or mounted onto a gearset, allowing the transmission axis orientation to be varied through the use of a thumbwheel. Because the polarizer is mounted in a sliding frame, it can be easily removed from the light path when not in use. The accompanying analyzer (a second polarizer necessary for fluorescence anisotropy measurements) can be mounted in the vertical illuminator above the filter block turret, or in a specially designed intermediate tube for polarized light microscopy. The auxiliary tube often enables insertion of a graduated, rotating polarizer unit to ensure accurate positioning of the analyzer transmission azimuth with respect to the polarizer and microscope optical axis.The last stage of the vertical illuminator contains a revolving turret or sliding bracket that houses optical blocks containing the fluorescence filter combinations. The filter blocks are rotated (or slid) into the optical train when required for imaging with the specific fluorescence filter combinations. Individual blocks contain a matched set of excitation and emission filters, as well as a dichromatic beamsplitter, all carefully positioned to maximize specimen illumination and capture of fluorescence emission at specific wavelengths. Optical block turrets and slider brackets can load between three and 6 individual blocks with one positioned in the light path, while the others are rotated or slid out of the way in temporary storage. These accessories enable rapid changeover between fluorescence filter sets when observing specimens labeled with two or more fluorescent probes. The use of transmitted illumination in a microscope equipped with a fluorescence illuminator is facilitated by dummy optical blocks that fit into a slot on the turret or slider bracket. The dummy blocks block excitation emission (when the shutter is opened), but do not contain any filters and allow light to pass unimpeded from the objective to the observation tubes.In standard modular upright microscopes, the vertical illuminator is positioned between the microscope frame and the observation tubes. Many manufacturers offer intermediate tubes that can be inserted between the illuminator and the eyepiece unit for polarizers, DIC prisms, or other accessories. Modern microscope frames are usually the result of computer-aided design efforts that result in significant vibration reduction and enhanced ergonomic features. Synthetic materials, such as metallic matrix ceramic and aluminum composites, dramatically improve the static and thermal rigidity of the frame, thus enabling the instrument to withstand the rigors of advanced fluorescence techniques that require an almost total elimination of vibration over the long periods of time necessary for imaging weak fluorescence emission.

Köhler Illumination in Fluorescence Microscopy

In a fluorescence vertical illuminator, the light source is positioned so that the filament or arc-discharge plasma ball is located near the principal focal point of the collector lens. In Köhler illumination, the lamp collector lens serves the function of a dramatically enlarged secondary light source to enhance overall illumination. One the primary requirements of Köhler illumination is that an image of the lamp filament or the arc must ultimately be projected onto the rear focal plane of the objective, which also doubles as the (often high numerical aperture) condenser during excitation in reflected light illumination. The light source should ideally fill the entire objective aperture to both maximize the intensity of radiation and to produce an evenly illuminated field. In some cases, a ground glass filter is placed into the vertical illuminator between the lamphouse and the neutral density filters in order to increase the uniformity of illumination, especially when employing arc lamps that tend to produce regions of excessive intensity (termed **hot spots**). However, because diffusion filters also reduce the level of illumination, they should be avoided whenever possible.

In reflected light Köhler illumination (illustrated schematically in Fig. 7.22), an image of the light source is focused by the collector lens onto the aperture iris diaphragm located in the vertical illuminator. This diaphragm shares a conjugate plane with the rear aperture of the objective and the lamp arc or filament, and therefore, determines the illuminated field aperture size. Together, the light source, vertical illuminator aperture diaphragm, and objective rear focal plane (pupil) form the illumination set of conjugate planes. Unlike the situation in transmitted light microscopy, the aperture iris and light source are imaged onto the objective (acting as a condenser) rear aperture plane, rather than being physically located at this position. As an added benefit to this configuration, all obstructions (such as iris diaphragms) are removed from the light path when the objective is providing excitation illumination or forming an image from the gathered fluorescence emission. Opening or closing the aperture diaphragm is used to control stray light and regulate the intensity (numerical aperture) of illumination without altering the size of the illuminated field. In the image, adjustment of the aperture diaphragm affects brightness and contrast.The image-forming or field set of conjugate planes in reflected light Köhler illumination consists of the field diaphragm, the specimen surface, and the intermediate image plane. Thus, when the field diaphragm is placed in focus at the specimen plane, the image of the light source is significantly removed from focus in order to provide a uniform field of illumination. The field diaphragm controls the size of the illuminated field without affecting the illumination intensity of the area being observed. In practice, the field diaphragm opening size should be as small as possible in order to increase image contrast and to reduce the degree of photobleaching in areas not being directly observed. Köhler illumination produces even illumination of the specimen field in spite of the uneven illumination intensity generated by most arc-discharge and filament-based light sources. When the microscope is properly configured, the rear focal plane of the objective is fully illuminated, providing a field that is uniformly bright from edge to edge. Köhler illumination, in the ideal case, bathes the specimen with a converging set of wavefronts, each arising from separate points on the light source imaged into the condenser aperture. In a properly configured fluorescence microscope, the result is optimum image contrast and resolution.

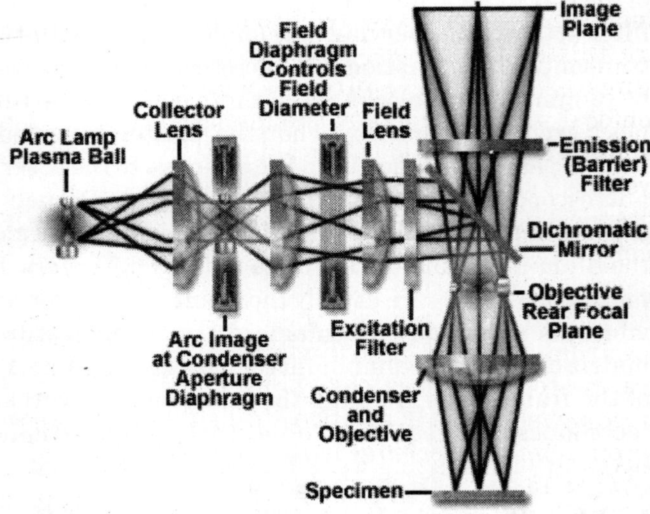

Fig. 7.22. Köhler Illumination in Reflected Light Fluorescence.

Light Sources and Lamphouses

For rigorous quantitative analysis in fluorescence microscopy, specimen illumination must be temporally and spatially constant over the entire viewfield. Instability over time generally reflects fluctuations of lamp radiance that occur as a result of electrical power supply output variations. In contrast, spatial disturbances, which occur most often in arc lamps, arise from a phenomenon known as **flicker** as the plasma ball migrates back and forth across the electrodes. Flicker is generally produced by power supply fluctuations, small variations in the electrode resistance, or mechanical vibrations. Filament-based light sources, such as the popular tungsten-halide lamp, are very stable when operated under constant direct current with a regulated power supply. In general, light sources should be chosen from their spectral content, filament size compared to the objective rear aperture area, spatial and temporal stability, and uniformity of field illumination. Careful attention must also be given to the intensity of the light source because the narrow band of wavelengths passed by the excitation filter comprises only a very small portion of the total illuminator output.The primary consideration in choosing a light source for fluorescence microscopy is the ultraviolet and visible light spectral distribution in relation to the quantum yield and absorption of fluorochromes being investigated. In addition, the source must be compatible with the sensitivity of the detector used of capture images, whether it is the human eye, traditional film, a photomultiplier, intensified video tube, or a digital camera system. The choice also depends on the mode of illumination. Widefield fluorescence microscopy requirements are fulfilled with arc-discharge or tungsten-halogen sources, while confocal and multiphoton microscopy requires the adaptation of various laser systems. Tungsten and tungsten-halogen incandescent lamps have been used with limited success due to the fact that a majority of their emission occurs in the red and infrared regions of the light spectrum, whereas most fluorophores are excited by ultraviolet, blue, and green wavelengths. In addition, the light output from arc-discharge lamps ranges between 10 and 100 times brighter than the 12-volt quartz halogen lamps typically used for transmitted light illumination. The favourite source of illumination for widefield fluorescence microscopy is the mercury arc lamp, which is usually routinely included in base-model microscope configurations. In some cases, xenon and metal halide arc lamps are employed, but these are usually confined to specialized circumstances where the light spectrum and intensity profile matches specific fluorophore requirements.

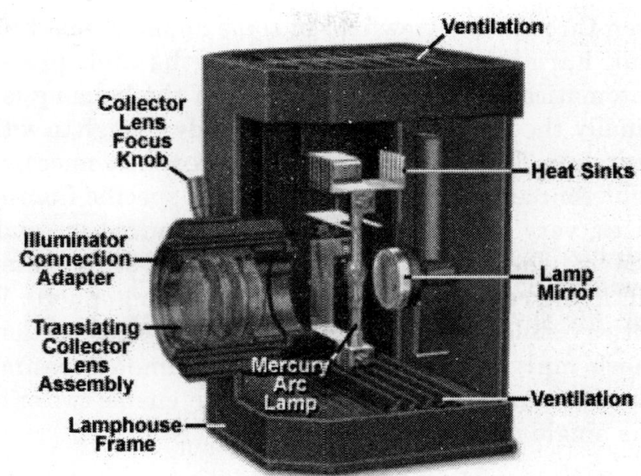

Fig. 7.23. Fluorescence Microscope Arc-Discharge Lamp Housing.

The design of mercury and xenon lamps is similar, except for the physical dimensions and the gas enclosed in the bulb envelope. Mercury lamps contain two electrodes sealed under high pressure in a quartz glass bulb that also contains vaporized elemental mercury. When the power supply is energized, a series of high voltage pulses is sent to the electrodes, which ionizes a small portion of the mercury gas, igniting the lamp. After firing, the voltage is reduced and the ionized gas serves to carry current and generate the plasma ball that forms between the two electrodes. During lamp operation, the vaporization of mercury generates a tremendous amount of heat and pressure inside the glass bulb, ultimately producing the high intensity light. Mercury arc lamps produce light that, even though continuous across the ultraviolet and visible spectrum, is concentrated in discrete wavelengths of 365, 400, 440, 546, and 580 nanometers. Xenon lamps have a more uniform intensity profile across the entire spectrum, from the ultraviolet to the infrared. The choice of fluorochromes is paramount in determining the proper light source for fluorescence microscopy. Some fluorescent probes have excitation bands that coincide with the prominent mercury lines, while others benefit from the more evenly distributed illumination of a xenon lamp.Proper alignment of arc lamps in fluorescence microscopy is critical in order to achieve Köhler illumination and to avoid bright and dark regions in the fluorescence image. As such, the quality of the lamphouse can often be judged by the stability of correct lamp alignment and by the efficiency of the adjustment knobs for maintaining the alignment. The lamp socket should be equipped with lamp centering screws to permit centering the arc image in the objective rear aperture, and the lamphouse should incorporate an infrared filter to block the very long wavelengths in the far red and infrared that generate a tremendous amount of heat. Several lamphouse designs have a built-in red suppression filter (for example, a Schott **BG38**), or a slot for such a filter, to eliminate a reddish background seen through the viewfield in some applications. In addition the lamphouse itself should not leak harmful ultraviolet wavelengths and, preferably, should incorporate a switch to automatically shut down the lamp if the housing is inadvertently opened during operation. Finally, the lamphouse should be sturdy enough to withstand a possible burner explosion during operation.The wide diversity of fluorescence microscopy applications often call for a range of light sources to meet the demands of specific fluorophores and imaging conditions. In some cases, very low irradiation may be required in combination with an ultra-sensitive camera system, whereas for other investigations, strong laser excitation may be necessary in order to kill living cells or selectively bleach a fluorophore. Wavelength requirements often span the entire visible region of the spectrum, as well as portions of the ultraviolet and infrared. Because these multiple illumination requirements cannot be met with a single light source, manufacturers now offer adapters that enable two or more lamps to be simultaneously attached to a single microscope.

Fluorescence Filter Combinations

As previously discussed, light passing through the lenses and diaphragms of the vertical illuminator finally encounters the excitation filter housed in an optical block positioned to coincide with the axial intersection between the illuminator light path and the microscope optical train. The excitation filter selects a narrow band of wavelengths from the wide spectrum generated by the lamp and passes them to the dichromatic mirror, which in turn, reflects the light through the objective and onto the specimen. Fluorescence emission gathered by the objective passes once again through the dichromatic mirror and the emission or barrier filter before forming an image in the fixed eyepiece diaphragm. The common filters employed to isolate wavelength bands in widefield fluorescence microscopy include colored glass filters and interference thin film filters (or a combination of the two). Identifying the proper filters for each step in the fluorescence microscopy illumination and imaging scheme can be confusing due the large number of filter manufacturers involved, each of which contributes a proprietary alphanumeric nomenclature to the fluorescence literature.The anatomy of a typical fluorescence filter block is diagrammed schematically in Fig. 7.9, along with the associated spectral profiles of the dichromatic mirror, excitation, and barrier filters. Filter blocks are usually assembled with custom tools supplied by the manufacturer so the operator can interchange filters and the dichromatic mirror. The excitation and barrier filters are secured into place with retainer clips, optical glue, or circular threaded mounts (see Fig. 7.9). In general, these filters can be removed without opening the optical block because they are positioned over recessed apertures on the flat exterior surfaces. Replacing a dichromatic mirror is more difficult and requires complete disassembly of the block to gain access to the interior. Most block sections are cast with a 45-degree diagonal joint that serves double duty by protecting the interior and supporting the dichromatic mirror at the proper angle. After removing the fasteners holding the block sections together (pins or small screws), the mirror can be removed by loosening or moving the retainer clip, and then very carefully dropping it out of the block. The dichromatic mirror should be handled with caution because the interference coating is usually not protected and can be easily scratched. Several of the filter manufacturers who supply the microscope companies also offer a wide selection of aftermarket filters and dichromatic mirrors for a variety of fluorescence applications.Fluorescence filter designs include longpass, shortpass (edge filters), and the narrow, medium, and wide family of bandpass filters. Examples of several common filter profiles are illustrated by the spectra presented in Fig. 7.9. The emission filter spectrum (blue curve) in Fig. 7.9 is produced by a longpass interference filter having a cut-on wavelength of approximately 575 nanometers. Longer wavelengths are transmitted through the filter, while shorter wavelengths are blocked. A narrow bandpass excitation filter from the same set (red curve, Fig. 7.9) has a bandwidth of approximately 20 nanometers, while the dichromatic mirror (green curve) has transmission regions approximating a medium (455-490 nanometers) and wide bandpass filter (560-775 nanometers). Because the dichromatic mirror effectively serves as a longpass filter in the green, yellow and red regions of the visible spectrum (560 to 700 nanometers), it is treated as such in filter sets. A working knowledge of how fluorophore absorption and emission spectral profiles can be utilized to select the appropriate filter set for fluorescence microscopy is essential to successful implementation of this technique.Dichromatic mirrors (or beamsplitters) are the most critical component in a fluorescence microscopy filter combination, and resemble longpass interference-type filters that are fabricated to close tolerances with multiple layers of dielectric materials. The major difference between a dichromatic mirror and a standard interference filter is that the mirror is specifically designed for reflection and transmission at defined boundary wavelengths, and must operate at a 45-degree angle with respect to the microscope and illuminator optical axes. Dichromatic mirrors are positioned with the interference coating facing the excitation light source in order to reflect short excitation wavelengths at a 90-degree angle through the optical train to the specimen. The same mirror must also act as a transmission filter to pass long wavelength fluorescence emission from the objective to the image plane. Because the wavelength transition region between almost total reflection and maximum transmission is often limited to only 20 or 30 nanometers, the dichromatic mirror is able to precisely discriminate between excitation and emission wavelengths. Fluorescence filter sets are designed so that a particular band of excitation wavelengths exactly matches the reflection region in the dichromatic mirror. The result is efficient reflection of excitation light through the microscope and onto the specimen. Fluorescence emission from the specimen must match a high transmission region in the dichromatic mirror in order to enable these wavelengths to pass through to the detector. The barrier filter is less important in the overall scheme, but still plays an important role in ensuring that scattered and reflected excitation wavelengths, fluorescence from probes other than the target, and general background intensity due to auto fluorescence is removed. The most important factor in creating a filter set is to make certain that the transmission, reflection, and emission profiles of the involved filters match in the appropriate regions. Otherwise, excitation wavelengths might be able to pass through the dichromatic mirror and fog the image, or fluorescence emission can inadvertently reflect at the mirror to compromise image brightness. Even in seemingly perfectly matched filter combinations, slight overlaps between spectral profiles of the individual filters can occur to diminish performance. Of particular concern is crosstalk between the excitation filter and the dichromatic mirror, which enables some of the excitation light to pass through the mirror and reflect from the walls of the filter block. Light reflected at highly oblique angles can be partially transmitted through the barrier filter, as discussed above, to reduce image contrast. This type of leakage through filters is termed **bleed-through** or **crossover**, and occurs to a greater or lesser extent with practically all filter combinations. One of the primary areas of concern with both filter manufacturers and microscope companies is to improve the design of fluorescence filter combinations in order to reduce the level of crossover.

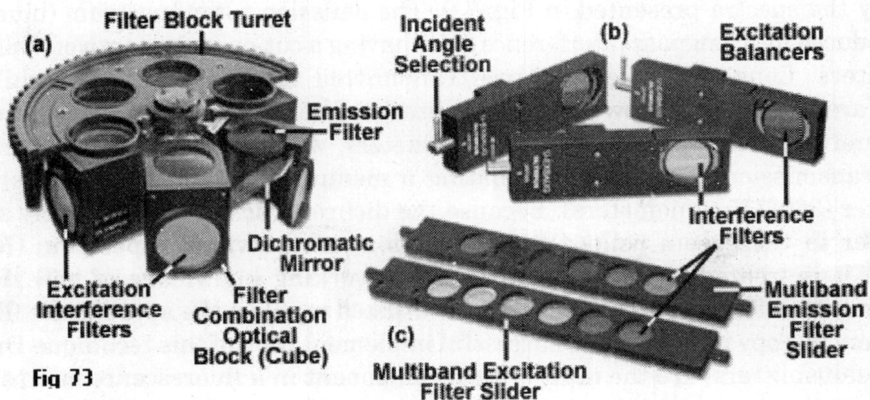

Fig. 7.24. Fluorescence Filter Set Configurations.

Several of the configurations for fluorescence filter sets are illustrated in Fig. 7.24. The filter block turret (Fig. 7.24(a)) contains five blocks, enabling simultaneous investigation of at least this number of fluorophores with rapid changeover between filter combinations. One of the turret slots is usually filled with a dummy optical block or left vacant for transmitted light observations. Fine-tuning of the excitation spectrum for imaging dual or multiply labeled specimens is accomplished with excitation balancers (Fig. 7.24(b)), which contain a single interference filter mounted on a swivel. Rotating the adjustment lever of an excitation balancer shifts the bandpass transmission region of the filter to shorter wavelengths. Thus, when the balancer filter is rotated from an incident angle of zero degrees (perpendicular to the vertical illuminator axis) to the maximum swivel angle of 45 degrees, the bandpass region of the filter can be shifted by values ranging between 25 and 50 nanometers. Excitation balancers can be employed singly or in tandem to alter the excitation bandwidths of fluorochromes under observation in order to equalize the emission intensities. This feature enables fine-tuning of fluorophore intensity in specimens containing several probes to, for example, reduce the fluorescence emission from one probe while increasing the intensity of another.

The rapid advances in thin film coating technology are evidenced by the creation of multiple transmission peaks in a single interference filter and the ability to fabricate interleaved bands of reflection and transmission in dichromatic mirrors. When properly matched, two multi-band filters and a dichromatic mirror can be combined to produce a multiple fluorescence filter set that enables simultaneous excitation and observation of emission by several fluorophores. Filter sets are available from the manufacturers that are suitable for use with two, three, and even four fluorophores in the same specimen. The major problem with multiple filter sets is their expense and the excessive amount of crossover or bleed-through that occurs when emission from one of the fluorescent probes is transmitted through a bandpass region intended for another. In some specimens (and filter sets), bleed-through has a significant impact on background intensity and image contrast. Many investigators prefer to image each fluorophore separately with optimized filter sets, subsequently combining the images into a composite.Advanced fluorescence techniques often require the use of several excitation and emission filters with a single dichromatic mirror. In order to facilitate these investigations, many fluorescence microscopes are equipped with motorized filter wheels or sliders containing up to six individual filters (see Fig. 7.24(c)). Microscopes designed to be used with the sliders either contain a special slot in the vertical illuminator for insertion of the slider, or the filter slider can be substituted for one that normally contains neutral density filters. Emission sliders can also be accommodated by the vertical illuminator in some microscope designs, or an auxiliary intermediate tube installed between the illuminator and observation tubes to house the slider. Motorized filter wheel units containing excitation filters are generally sandwiched into the vertical illuminator optical pathway between the neutral density filters and the lamphouse in both upright and inverted microscopes. Likewise, the motorized emission filter wheel unit is placed between the vertical illuminator and the eyepiece observation tubes in upright microscope designs, but is usually attached through the same external port as the detector for inverted instruments. Excitation and emission sliders and motorized filter wheels are extremely useful for dual and triple excitation applications with a single dichromatic mirror. By eliminating the need to re-position filter blocks while observing a single specimen, these accessories enable the investigator to obtain accurate registration between images.

Objectives for Fluorescence Microscopy

In all forms of reflected light microscopy (including fluorescence), image intensity is a function of the objective numerical aperture and magnification. In fact, the intensity or brightness (defined as photon flux per unit area and time) increases by the fourth power of the numerical aperture, but is inversely proportional to only the square of the magnification:

Intensity µ $(NA_{obj})^4/M^2$

where **NA** is the objective numerical aperture and **M** is the magnification. It is obvious from this relationship that the brightest fluorescence images will be gathered by an objective with high numerical aperture and low magnification (such as a 0.75/20x). As an example, a 60x plan apochromat oil immersion objective with a numerical aperture of 1.4 will theoretically produce brighter images than a 100x objective of the same numerical aperture, but there are trade-offs to consider. Increasing the number of internal lens elements (as with the highest numerical aperture objectives) produces corresponding increases in autofluorescence and intensity-reducing reflections from the internal lens surfaces. Often, a compromise between the highest correction factors and increased light transmission is made by the manufacturers in their recommended objectives for fluorescence microscopy.In general, the high numerical aperture oil (1.3 to 1.4) and water (1.2) immersion plan fluorite and plan apochromatic objectives produce the brightest fluorescence images because of their tremendous light-gathering ability. These objectives exhibit excellent color correction, and thus, are able to focus a wide range of fluorescence emission wavelengths in the same plane. The light transmission characteristics of apochromatic and fluorite objectives are excellent down to about 350 nanometers, an absolute requirement for examining fluorochromes that are excited in the ultraviolet region (such as DAPI, Hoechst, Alexa Fluor-350, and AMC). Despite having numerous internal lenses, these objectives are fabricated from low-fluorescence glass with anti-reflection coatings to minimize background fluorescence and yield images of very high contrast. Image brightness during observation can also be increased by reducing the eyepiece magnification factor from the standard 10x down to 8x or lower.Objectives designed for specialized applications are widely available for fluorescence microscopy. Included in this class are high numerical aperture water immersion, multi-immersion (oil, water, and glycerin), and ultraviolet objectives that have quartz lens elements. For live cell imaging applications, long working distance objectives with correction collars are necessary for viewing specimens through thick (0.5-2 millimeter) culture flask walls. Examinations deep within thick specimens are facilitated by objectives having a very long focal length (working distance), which are available from several manufacturers for imaging with cover glasses through air or water. Long working distance water immersion objectives designed to be used without a cover glass also have a Teflon nosecone so that the objective can be dipped into aqueous solutions. Similar objectives are produced for ultraviolet excitation (340 nanometers) in water at working distances between 20 and 30 millimeters (including approximately 5 millimeters of water). For relatively low magnification investigations, 20x water immersion objectives with very high numerical aperture (0.75 to 0.95) are recommended. Although quite expensive, these objectives produce extremely bright images in situations where fluorophore concentrations and/or quantum yields are marginal.

Accessories for Fluorescence Microscopy

Manufacturers are continually producing useful add-on accessory units for their instruments in order to increase the available options for the ever-growing number of imaging applications in fluorescence microscopy. For example, a rectangular field stop (see Fig. 7.25), which restricts the specimen area being illuminated to improve contrast and reduce photobleaching, is available as an option on some microscopes. The modular field stop is substituted for the traditional iris field diaphragm (also a removable module) in the vertical illuminator and is designed to match the aspect ratio of digital imaging sensors. Rectangular field stops improve the efficiency of Köhler illumination and reduce the signal-to-noise ratio for electronic sensors. As an added benefit, the image rectangle size can be adjusted to coincide with stage step size when using programmable scanning stages in deconvolution investigations. Pinhole apertures, housed in modules similar to the rectangular field stop, are available for applications requiring a highly restricted illumination field.

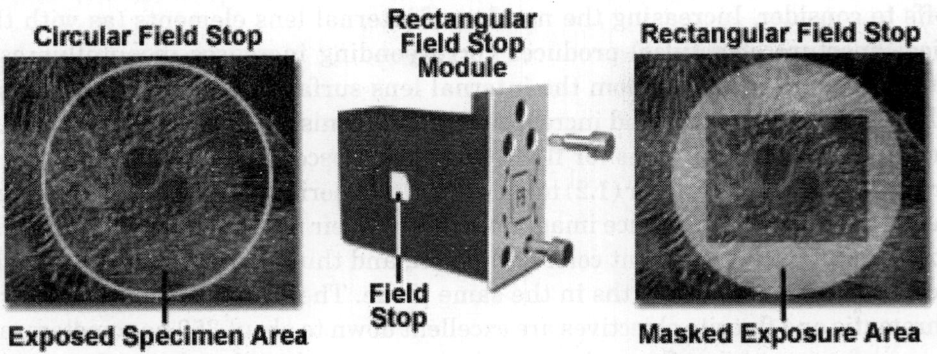

Fig. 7.25. Circular and Rectangular Field Stop Comparison.

Other accessories include double lamp housing adapters that enable two light sources (such as mercury and xenon arc-discharge lamps) to be simultaneously attached to the vertical illuminator. Specialized illuminators that adapt lasers for excitation are also available for applications such as total internal reflection fluorescence (**TIRF**), fluorescence lifetime imaging microscopy (**FLIM**), ratio imaging, and photobleaching recovery experiments. In addition, most of the research-level fluorescence microscopes offered by the major manufacturers are readily adapted to their respective laser scanning confocal accessories. Upright microscopes accept confocal scan units through the trinocular observation tube, while inverted instruments can attach the beam scanners to side or rear ports in the microscope frame. Dual ports that are inserted between the vertical illuminator and the observation tubes can be useful for both introducing additional light sources or directing fluorescence to more than one detector. Most of these units are available with optional C-mount adaptors or standard microscope port fittings. Dual port adapters are also available for simultaneously attaching a fiber optic cable from a laser source and an arc-discharge lamphouse to the rear of the vertical illuminator.Fluorescence microscopes designed for electrophysiology investigations have become very sophisticated. Many are equipped with specialized vibration-dampened stages that can accept a wide variety of micromanipulator accessories for experiments using fluorescence, infrared differential interference contrast (**IR-DIC**), and traditional brightfield contrast-enhancement imaging techniques. Swinging nosepieces enable rapid vibration-free interchange of objectives to prevent disturbing specimens or the intrusion of air bubbles while imaging live cells and tissues. Fluorescence macro observation in living organisms is possible with new high numerical aperture long working distance 2x and 4x macro lenses (objectives), which are equipped with specialized filter combination blocks. In addition, several manufacturers offer fluorescence vertical illuminators and filter sets as accessories for their stereomicroscopes.Magnification factors of 1.25x to 1.5x are sometimes available as an option on vertical illuminators, but the use of auxiliary lenses should be avoided whenever possible due to the inherent problems of introducing **empty magnification** into fluorescence images. Several modular vertical illuminator designs include provisions for attaching beamsplitter modules containing multiple camera ports above the illuminator to increase imaging capabilities. Considering the wide range of motorized accessories now available for fluorescence microscopes, including nosepieces, condensers, filter wheels, and stages, more sophisticated imaging techniques are available to investigators than ever before. The manufacturers exert a significant amount of effort in the design of accessories for many of the complex fluorescence applications that were once possible only with microscopes constructed using aftermarket parts from a variety of sources. As fluorescence microscopy becomes an increasingly critical tool in cell biology, neurophysiology, and the clinical arena, the development of innovative new microscope accessories will no doubt be a continuing trend.

Inverted Fluorescence Microscope Design

A similar version of the vertical fluorescence illuminator is available for inverted (tissue culture) microscope stands. The inverted stands also permit combining or alternating between reflected light fluorescence and the various contrast enhancing techniques of transmitted light microscopy. Research-level inverted microscopes feature multiple (up to six) input/output ports, usually with single ports on each side of the frame, as well as one or two ports (upper and lower) at the rear and a bottom port underneath the base of the microscope. In some models, primary images can be obtained simultaneously from three or more ports without the use of relay lenses. This level of connectivity enables the use of multiple light sources, filter wheels, and camera systems for complex fluorescence analysis. Mercury and xenon lamphouses for inverted microscopes are available with the standard multi-element or aspherical collector lenses to improve performance and reduce aberration in the ultraviolet and infrared spectral regions. In addition, a wide spectrum of lamphouse adapters can be utilized to attach several illumination sources, similar to the accessories available for upright microscopes.

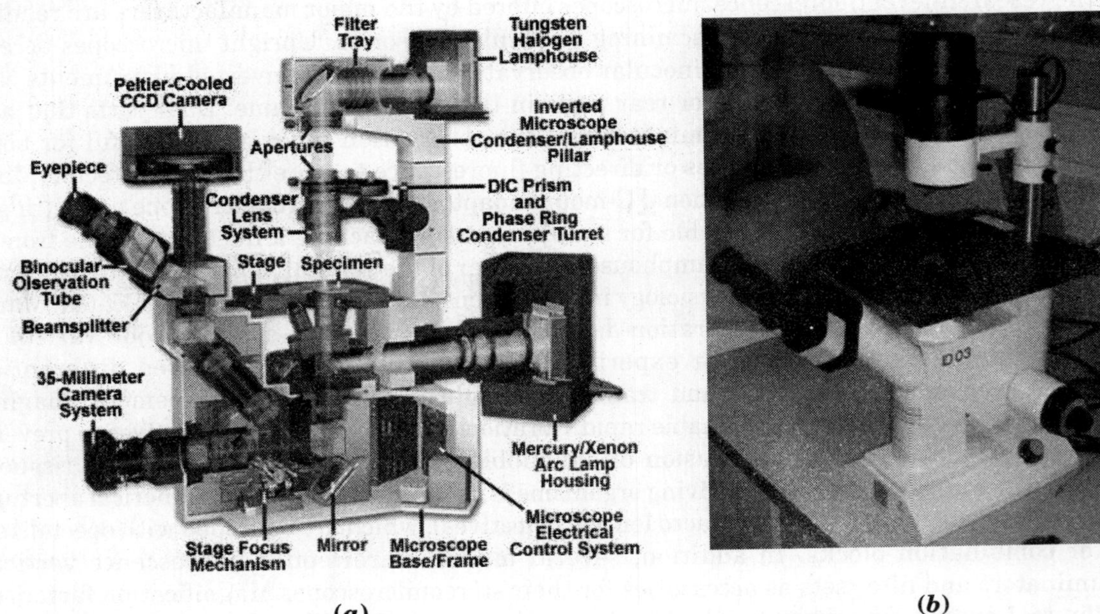

Fig. 7.26. The Inverted Fluorescent Tissue Culture Microscope the Internal Structure and the real Picture.

Presented in Fig. 7.26 is a cut-away schematic diagram and the actual photograph of a modern inverted (tissue culture) fluorescence microscope equipped with both a Peltier-cooled CCD image sensor and a traditional 35-millimeter film camera system. Although film cameras are of limited use, most inverted microscopes still include a port for these attachments in the lower portion of the front base. The microscope illustrated in Fig. 7.26 can perform traditional brightfield transmitted illumination (with or without contrast enhancement) using the tungsten-halogen lamphouse mounted on the pillar. A mercury or xenon arc-discharge lamp is employed for fluorescence microscopy using a reflected light illuminator attached to a specially configured rear port. Aperture and field diaphragms, along with neutral density filters, are accessed near the port at the rear of the microscope. In some inverted microscope models (not illustrated), an L-shaped reflected light illuminator containing centerable diaphragms is available to improve access to the rear auxiliary ports for additional accessories. The light path through the microscope in Figure 75 is depicted in yellow for transmitted light, violet for unfiltered arc lamp illumination, green for filtered fluorescence excitation, and red for fluorescence emission.Modern inverted microscope frames, like their upright counterparts, are computer engineered and fabricated with composite materials for structural and thermal stability. In addition, the mechanical stage components and circuit structures are designed with short travel distances and high rigidity to avoid pitch and yaw when the nosepiece is manipulated during routine operations such as DIC prism insertion or adjustment of objective correction collars. Advanced nosepiece stages are able to virtually eliminate focus drift during time-lapse and prolonged fluorescence observations. Other stage options not available for standard upright microscopes include gliding and 360-degree rotating stages, glass stage insert plates, heating plates, Petri dish and plate holders, and carbon dioxide incubator chambers. Inverted microscopes having a modular design can easily be configured for investigations in electrophysiology, ***in vitro*** fertilization, micromanipulation, high-resolution DIC, video-enhanced observations, and a variety of advanced fluorescence techniques. The instruments are also readily adapted for confocal and multiphoton microscopy. Motorized accessories include shutters, filter wheels, revolving nosepieces, fluorescence block turrets, focus drives, and condensers. When coupled to the advanced objectives available in long working distance, water immersion, ultraviolet excitation, and phase contrast, all having a high degree of optical correction, inverted microscopes are ideal instruments for conducting fluorescence investigations on living cells and tissues.

Conclusions

In fluorescence microscopy, wide variations between localized fluorophore concentrations within the specimen, coupled to differences in extinction coefficient and quantum yield from one fluorochrome to another, significantly influence the emission signal produced for a given quantity of excitation intensity. Considering that many specimens contain only minute quantities of fluorescent material in any particular viewfield, these combined factors produce an average level of fluorescence emission that is four to six orders of magnitude less than the excitation intensity. In addition, several of the more sophisticated fluorescence techniques, such as ***in situ*** hybridization and resonance energy transfer (**FRET**), have emission signal intensities that can range nine to ten orders of magnitude less than that of the excitation. In order to compensate for these large discrepancies between the intensity of excitation and emission, modern fluorescence microscopes must be able to attenuate the excitation illumination by levels exceeding a billion times without disturbing the fluorescence signal.Among the main attributes of fluorescence microscopy is the high specificity for fluorescent probes that absorb and emit light at characteristic wavelengths, leading to the ability of the technique to selectively detect a target species at very low concentrations in complex mixtures. In addition, the high sensitivity and spatial resolution of fluorescence enables single molecules to be pinpointed and studied at length scales beneath the optical resolution of the microscope. Localized environmental factors also weigh heavily on fluorescence emission, so the method is an ideal probe for fluctuations in pH, viscosity, ion concentrations, molecular distances and orientation, membrane potentials, hydrophobicity, charge distribution, and diffusion coefficients. The temporal resolution of fluorescence is limited to the lifetime of excited fluorescent probes, which can be on the order of nanoseconds. Because many biological processes occur in this time domain, decay kinetics can reveal dynamic information about cellular processes. Taken collectively, these factors are of critical importance in applications of fluorescence microscopy to cell biology. Fluorescence microscopes have evolved with amazing speed over the past decade, coupled to equally rapid advances in laser technology, solid-state detectors, interference thin film fabrication, and computer-based image analysis. The development of high numerical aperture water immersion objectives has further assisted investigations of biological phenomena, allowing investigators to probe deep within the living cell in its natural environment. As the microscope manufacturers respond to the changing needs of the research community, the development of advanced fluorescence instruments and accessories will no doubt continue, and their ultimate contribution to explorations into nature's mysteries may ultimately be of profound significance.

Troubleshooting Photomicrography Errors (Fig. 7.27)

Photomicrography under fluorescence illumination conditions presents a unique set of circumstances posing special problems for the microscopist. Exposure times are often exceedingly long (in some instances running from many seconds into several minutes), the specimen's fluorescence may fade during exposure, and totally black backgrounds often inadvertently signal light meters to suggest overexposure.

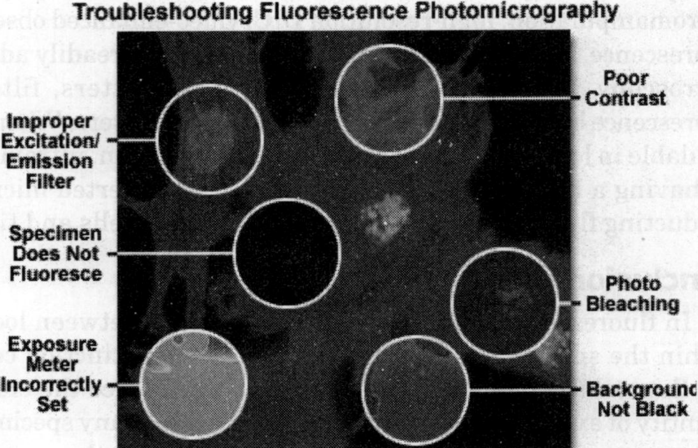

Fig. 7.27. Troubleshooting Fluorescence Photomicrography.

In addition, fluorescing specimens emit their own light, and particles residing above and below the desired plane of focus often radiate light causing blurring of the image details. Even though fluorescence images may appear to be bright when viewed through the microscope eyepieces (due to the human eye's exquisite sensitivity to light), they usually require lengthy exposure times in order to register a satisfactory image on film. Very long exposures complicate photomicrography of specimens stained with fluorescent dyes by increasing the likelihood of vibration and film reciprocity failure effects causing undesirable color shifts in the image.

While the basic problem with fluorescence photomicrography is often the paucity of light reaching the film, there are a number of steps that can be taken to improve image quality. Microscope objectives utilized for fluorescence should have as high a numerical aperture as possible, with high transmission values for the wavelengths of light being employed in the experiments. Because the light intensity (in reflected light fluorescence) varies as the fourth power of the numerical aperture, these objectives can significantly reduce exposure times in reflected light fluorescence microscopy. Light intensity also varies inversely as the square of the magnification, resulting in brighter images when the total magnification on film is held to a minimum. Reduction of magnification on the film plane is often assisted by the use of low magnification photoeyepieces (or projection lenses). In addition, when using a trinocular head, it is usually desirable to divert all light to the camera rather than have the light split (by means of a beamsplitter) between the viewing eyepieces and the camera. Because the image intensity also varies inversely as the square of the distance to the film plane, microscopes perform better then this distance is kept short. For this reason, and due to widespread availability, most fluorescence microscopy is recorded on 35 millimeter film (as opposed to larger format) when using traditional photomicrography to capture images. Modern low-noise temperature-controlled CCD digital cameras are being increasingly used to replace film as the medium of choice for fluorescence photomicrography.

Low light levels and the fading of fluorescent specimens are two of the primary difficulties to overcome in fluorescence photomicrography. Often, the dynamic nature of fluorescent preparations makes good photomicrography essential to capture a record of events occurring with the specimen, and often may reveal details not obvious when examining the specimen through the eyepieces. Long exposures, so commonly used in fluorescence, inevitably lead to reciprocity failure and fading of specimen fluorescence intensity during prolonged irradiation. In fact, fading of specific fluorescent features is often faster than that of the background, leading to a loss of image contrast.

All photographic emulsions (including black and white, colour negative, and color transparency films) suffer from an effect known as reciprocity law failure when exposed to light for exceeding long or short time periods. The reciprocity law relates image density to exposure time and light intensity, such that there is a reciprocal relationship between illumination intensity in the microscope and the required exposure time for film. In effect, the law states that exposure time is inversely proportional to light intensity in order to maintain a constant film density. The lower the light intensity, the longer the necessary exposure time required to produce a satisfactory image. This inverse relationship holds true for most films if the proper exposure duration is between 1/500 of a second and 1/2 second.

In fluorescence microscopy, low light levels routinely demand exposures that exceed 1/2 second (often ranging from several seconds to several minutes). The reciprocity relationship no longer holds under these conditions, and the film will require additional exposure time to yield the proper image density. This phenomenon is called reciprocity failure, a term that only indicates that the linear relationship between exposure time and light intensity no longer holds and does not suggest a failure of the film emulsion in terms of performance.

A side effect of reciprocity law failure is that the already weak specimen fluorescence is further reduced by photobleaching of the specimen, and incorrect color reproduction often occurs due to increased exposure times. The series of images presented in Figure 77 are fluorescence photomicrographs illustrating antibody fluorochrome labeling with the popular reagent fluorescein-5-isothiocyanate (FITC) when imaged using an Olympus WIB long-pass fluorescence cube. The cells in the center (Figure 7.28(b)) were photographed using a corrected exposure time adjusted for reciprocity law failure and Kodak Wratten color compensating filters to remove an undesirable color shift. On the left (Figure 7.28(a)), the same viewfield shows the effect of photobleaching, or fading of the fluorescence activity due to prolonged exposure to the illumination source. Color shifts and reduced image intensity resulting from reciprocity law failure are evident in Figure 7.28(c).

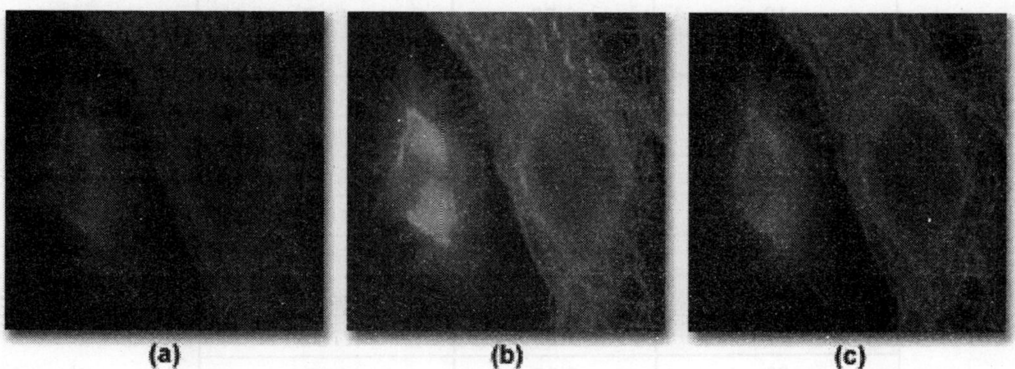
(a) (b) (c)

Fig. 7.28. Photobleaching and Reciprocity Failure in Fluorescence Photomicrography.

Reciprocity failure can often be compensated simply by an increase in exposure time or processing conditions for black and white films, but this is not always the case with color negative and transparency films. Most colour films have three color-sensitive dye layers, each of which has a slightly different characteristic curve position and slope resulting in varied responses to the reciprocity effect with the potential to cause undesirable color shifts or casts. Often, both exposure times and color balance filters must be adjusted to compensate for very long or short exposures when using color films.

Utilization of an objective having a very small numerical aperture (for instance, a 40x objective with NA = 0.65 as opposed to the same magnification objective having a NA of 0.85, 0.95, or 1.0) produces an image displaying insufficient brightness, which results in longer exposure times. The numerical aperture is a measure of the objective's light gathering ability, meaning that high numerical aperture objectives can capture a greater amount of the weak fluorescence emission produced by chromophores. Always use the highest available objective numerical aperture when doing fluorescence photomicrography. Using excessive objective and photoeyepiece magnifications will also result in reduced brightness. To compensate for this, always employ the lowest possible magnification photoeyepieces and the highest numerical aperture objectives at a given magnification. For example, a 60x plan apochromat objective having a numerical aperture of 0.95 (brightness index = 22.6) is preferable to a 60x plan fluorite objective of numerical aperture equal to 0.85 (brightness index = 14.5) even though the working distance of the fluorite objective is twice that of the apochromat. Furthermore, since high numerical aperture objectives have a shallower depth of field, their use may avoid contrast-reducing effects of out-of-focus emission from above or below the plane of focus. Likewise, a 2.5x photoeyepiece will provide more illumination than a 3.3x or 5x eyepiece. Laser scanning confocal microscopes are specifically designed to virtually eliminate out-of-focus light and give much clearer fluorescence pictures with somewhat better resolution.

Table 7.4. Objective Brightness Values.

Objective Magnification	*Numerical Aperture (NA)*	*Objective Brightness Index*
10	0.25	3.9
10	0.30	8.1
10	0.45	41.0
10	0.50	62.5
20	0.40	6.4
20	0.50	15.6
20	0.75	79.1
40	0.65	11.1
40	0.75	19.8
40	0.95	50.9
40	1.00	62.5
40	1.30	178.5
60	0.85	14.5
60	0.95	22.6
60	1.40	106.7
100	1.25	24.4
100	1.40	38.4

Typical fluorescence applications in fields such as cell biology and medical diagnostics are not only concerned with specimen brightness, but also the ratio of specific specimen fluorescence to the intrinsic background fluorescence. This is particularly important for single-molecule fluorescence events and other low light experiments where autofluorescence and/or internal reflection inside the objective produces vast differences in the ability to image small structures with low fluorescence quantum yields. To maximize the signal-to-noise ratio (intensity of the specimen fluorescence above the background) in objectives designed for fluorescence microscopy, manufacturers use quartz and other special glass formulations that have high transmission values throughout the spectrum from infrared to the ultraviolet. These objectives are extremely low in autofluorescence and utilize special optical cements and anti-reflection coatings designed to operate through an extended range of fluorescence excitation wavelengths. Higher numerical apertures coupled to advanced glass formulations provide modern fluorescence objectives with the ability to excite specimens at wavelengths down to 340 nanometers with much higher transmission values in order to obtain a brighter fluorescence image.

A comparison of nominal objective brightness values for common magnifications and numerical apertures is presented in Table 7.4. These numbers were calculated according to the following equation:

$$\textbf{Brightness Index} = (\textbf{NA})^4 / (\textbf{Magnification})^2 \times \textbf{Transmission Ratio}$$

where **NA** is the objective numerical aperture and the **Transmission Ratio** is presumed to be 100,000.

From this equation, it is apparent that for the same magnification, image brightness of both the illumination field and secondary fluorescence increases with the numerical aperture of the objective, but also decreases with increasing magnification. The actual numerical aperture and magnification values for a particular objective may vary by as much as 5-10 per cent from the nominal values. Brightness transmitted by an objective, as discussed above, is also dependent upon the internal construction parameters, such as the number of glass lens elements, internal reflections and glare, and lens coatings. Often, it is necessary to sacrifice higher optical correction of chromatic aberration found in apochromat objectives for the enhanced image brightness displayed by higher numerical aperture fluorite objectives with lesser correction.

Use as high a numerical aperture condenser as possible with transmitted light fluorescence to avoid light loss and the accompanying long exposure times. Also when using a substage mirror on older microscopes, purchase a mirror that has an aluminized surface instead of a silvered mirror (silver will reflect ultraviolet light poorly). Often, an oil immersion darkfield condenser may be substituted for standard brightfield condensers for transmitted fluorescence applications. When using immersion oil, apply a non-fluorescent oil between the top lens of the condenser and underside of the microscope slide and between the objective front lens and the coverslip. Oils of this type are readily available from a number of commercial manufacturers.

Photobleaching, or dye photolysis, rapidly degrades fluorescent probes used to stain specimens and should be minimized by exposing the preparation to the lowest illumination levels as is possible. This effect is caused primarily by the photodynamic interaction between the fluorochrome and oxygen, which involves promoting the dye molecule from the singlet ground state to the relatively long-lived triplet excited state by a process termed **intersystem crossing**. Once the chromophore has been elevated to an excited state, it becomes more chemically reactive and may then participate in irreversible chemical reactions including decomposition, polymerization, oxidation (primarily by singlet oxygen) or reaction with another molecule. Reaction with oxygen will usually cause bleaching of the chromophore, depending upon the intracellular singlet oxygen concentration and the proximity of the chromophore to other internal cellular components such as proteins, lipids, or small molecules. Calculations indicate that singlet oxygen can produce chromophore photolysis over a distance exceeding 500 angstroms.

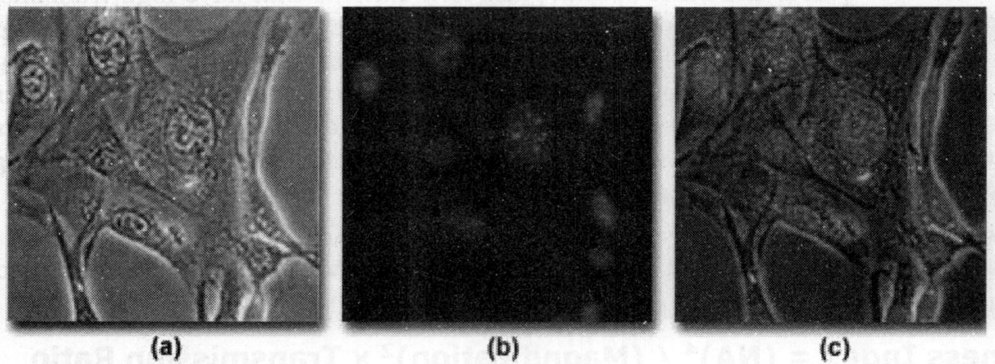

Fig. 7.29. Combined Use of Phase Contrast and Fluorescence Illumination.

To minimize the effects of photobleaching, fluorescence microscopy can be combined with other techniques that are non-destructive to the fluorochromes, such as differential interference contrast (DIC), Hoffman modulation contrast (HMC), transmitted darkfield illumination, and phase contrast. The idea is to locate the specific area of interest in a specimen using the non-destructive contrast enhancing technique then, without relocating the specimen, switch the microscope to fluorescence mode. The results of a typical experiment of this type are illustrated in Figures 7.29 and 7.30. Figure 7.29(a) illustrates 3T3 fibroblasts in monolayer tissue culture imaged using phase contrast optics. The cell line was established from a National Institutes of Health line of Swiss mouse embryo cells, which are highly contact inhibited and useful for studies involving sarcoma virus formation and leukemia virus propagation. The photomicrograph in Figure 7.29(b) shows the same viewfield, but this time imaged using fluorescence illumination (a

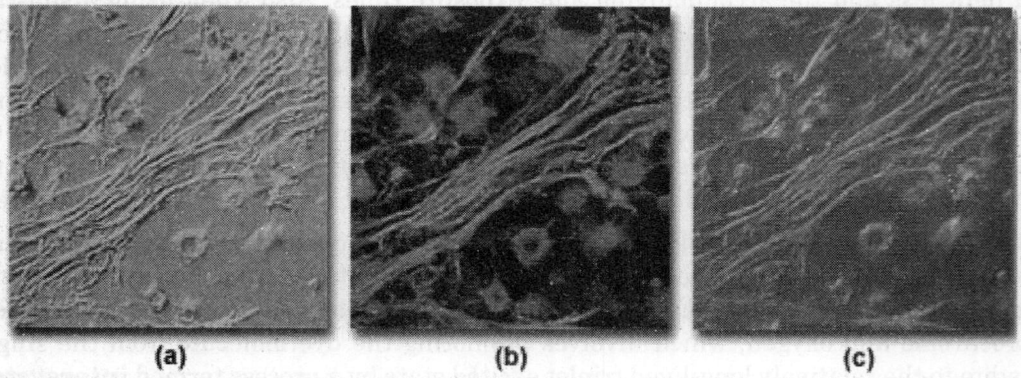

Fig. 7.30. Combined Use of DIC and Fluorescence Illumination.

mercury vapor lamp and an Olympus WU filter cube) with cells stained by the fluorochrome 4′,6-diamidino-2-phenylindole (DAPI), a nucleic acid specific dye with an emission maximum at 461 nanometers, which is used to selectively stain nuclei and chromatin. Figure 7.29(c) illustrates the two techniques used in combination to produce a beautiful photomicrograph of fluorescent-stained 3T3 cellular nuclei superimposed on a phase contrast image of the fibroblast cell membranes and internal organelles. This image was recorded with a specialized fluorite objective designed with phase rings to permit simultaneous observation of both fluorescence and phase contrast with the same objective.

Another example of combined techniques, this time using differential interference contrast (DIC), is presented in Fig. 7.30. Illustrated in Figure 7.30(a) is a thin section of cat brain tissue infected with cryptococcus and imaged using DIC optics and a full-wave retardation plate. Note the pseudo three-dimensional appearance of the photomicrograph. The cells in Figure 7.30(b) were stained with a combination of fluorescein-5-isothiocyanate (FITC) and Congo red (emission wavelength maxima of 520 and 614 nanometers, respectively). The two techniques are used in combination to produce Figure 7.30(c), which illustrates the infected cat brain tissue in both fluorescence and DIC illumination.

The rate of photobleaching is dependent upon several factors, including the chemical reactivity of the chromophore, the intracellular chemical environment, and the intensity and wavelength of the excitation light. Some fluorescent dyes are readily susceptible to photobleaching while others are relatively insensitive and are stable for much longer periods. In many instances, the specimen may recover from the effects of photobleaching, particularly if it is kept cool and in a dark environment. Photobleaching should be distinguished from another fluorescence artifact termed **quenching**, which occurs by reduction (or in some cases, enhancement) of fluorescence intensity by competing processes such as temperature, high oxygen concentrations, and molecular aggregation in the presence of salts or halogen compounds. Sometimes quenching results from the transfer of energy to other acceptor molecules residing physically close to the excited fluorochromes, a phenomenon known as resonance energy transfer. This particular phenomenon has become the basis for a newer technique of measuring distances far below the lateral resolution of the light microscope. Impurities in the chromophore may also reduce fluorescence intensity either by photobleaching or quenching.

Photobleaching can be minimized by reducing exposure times (as discussed above) or by lowering the excitation energy (lamp intensity), however these remedies are accompanied by the undesirable effect of reducing the chromophore fluorescence emission energy. Neutral density filters can be placed in the light path before the light reaches the excitation filter, thus diminishing the excitation light intensity and reducing fading effects during specimen observation. Many fluorescence microscopes are equipped with a shutter system coupled to neutral density filters that allow excitation light to be reduced in intensity during observation and specimen manipulation, but enable the microscopist to easily switch to full-power illumination for photomicrography. It is also helpful, where possible, to deoxygenate specimens (but not living cells or tissues) and to use specific antifade reagents such as ***n***-propylgallate and other inhibitors, which are commercially available. Chemicals that are capable of quenching singlet oxygen can also be employed to reduce the effects of photobleaching. Examples are 2-mercaptoethylamine, 1,4-diazabicyclo-2,2,2-octane (DABCO), diphenylisobenzofuran, ***p***-phenylenediamine, and the amino acid histidine. Fading effects can also be reduced, in some instances, by changing the pH concentration of the mounting medium.

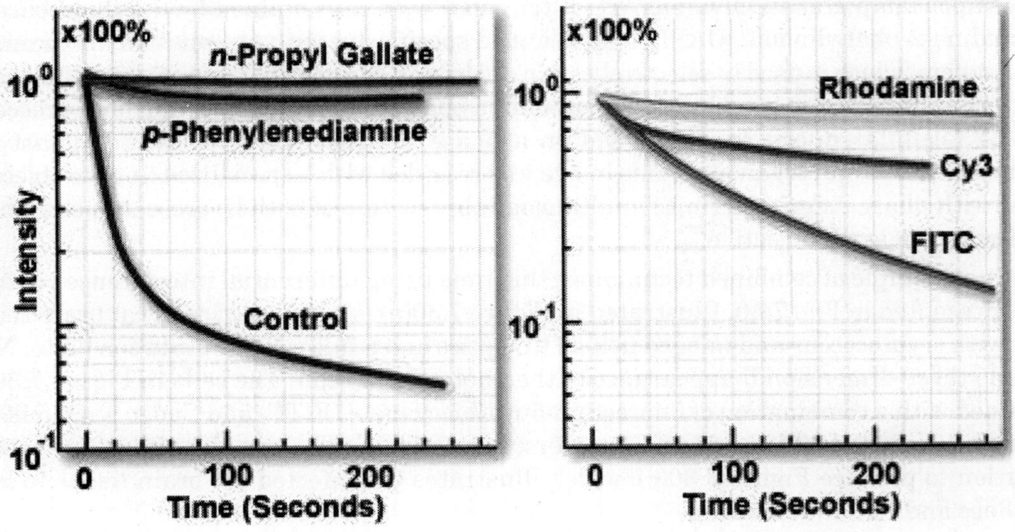

Fig. 7.31. Inhibition of Photobleaching and Specimen Fading.

Figure 7.31(a) illustrates inhibition of the rate of fluorescence fading by several commercially available antifade reagents. The Fig. shows the rapid decline in fluorescence intensity with an untreated specimen, demonstrate reduction of the fading rate by two antifade reagents of slightly differing efficiency. The specimen is marsupial kidney tissue culture cells (PtK2) stained with either fluorescein-5-isothiocyanate (FITC) or rhodamine phalloidin (both dyes have similar fading properties). Antifade reagents were ***p***-phenylenediamine or ***n***-propyl gallate. Specimen secondary fluorescence fading rates usually differ depending upon the fluorochrome in use, even with observation under identical conditions. The best results can be obtained by selecting a fluorochrome with the lowest fading speed from those available for a particular application. This concept is illustrated in Figure 7.31(b), which compares the difference in marsupial kidney cell fading speed for three fluorochromes: rhodamine, the cyanine dye Cy3, and FITC. Rhodamine, a ubiquitous dye with many applications, displays a very slow fading rate (Figure 7.31(b); under these conditions and is the best fluorochrome for photomicrography using this system.

To further minimize fading, it is advisable to conduct initial observations in one field within the specimen and then to quickly move to a "fresh" field just prior to taking the exposure. This practice may help to circumvent bleaching and/or fading effects. It is also helpful to do both observation and photomicrography in a partially darkened room environment. Although chemical methods and careful microscopy can reduce photobleaching, the most effective remedy is to increase detection sensitivity (coupled with reduced excitation energy) using a low-light level CCD digital camera designed specifically for fluorescence microscopy.

The occurrence of photobleaching has led to a technique known as **FRAP**, which is an acronym for Fluorescence Recovery After Photobleaching. This technique is based upon bleaching by short laser bursts and subsequent observation of the recovery of fluorescence caused by the diffusion of fluorochromes into the bleached area.

Barrier filters are designed to absorb specific wavelengths of excitation radiation that are reflected from, or transmitted by, the specimen and to transmit only the chosen visible light fluorescence. The cutoff wavelength range of the barrier filter is critical, because in many instances it must be able to transmit visible wavelengths that are not far in value from the excitation light used to illuminate the specimen. Because film is very sensitive to ultraviolet and visible wavelengths in the 380-450 nanometer region, it is particularly important that the barrier filter block all unwanted wavelengths in this range. To ensure removal of shorter wavelengths, additional ultraviolet blocking filters, such as the Kodak Wratten filter numbers 2A, 2B or 2E, should be added to the optical pathway. All of these filters completely absorb ultraviolet light, but have a slight difference in their absorption of lower wavelength visible blue light. Supplementary ultraviolet filters should be placed before the main barrier filter in the light path in order to block radiation that might induce autofluorescence in the main barrier filter, especially if this filter is yellow, orange, or red in color. If the barrier filter cuts the spectrum at too long a wavelength, specimen fluorescence color will be affected and the intensity of the image may be compromised, especially when only the long-wavelength portions of transmitted fluorescence are passed by the filter.

Colour balance errors in fluorescence photomicrography may have a number of origins. The photomicrographs presented in Fig. 7.32 illustrate several problems that arise with colour degradation from autofluorescence and improper filtration when imaging specimens under fluorescence illumination. The specimen is a monolayer of fibroblast cells grown in tissue culture and stained with 7-amino-4-methylcoumarin-3-acetic acid (AMCA), a bright ultraviolet-excitable dye conjugate with an emission wavelength maximum of 445 nanometers. The optimal excitation wavelength of AMCA is 345 nanometers, which requires an excitation filter that passes a significant amount of ultraviolet light. Under proper conditions of illumination and filtration, the stained specimen appears with nuclei and cytoplasmic components stained a rich and deep blue color (imaged with an Olympus WU filter cube), as illustrated in Fig. 7.32(b). When a barrier filter that passes too much ultraviolet light is used, a blue cast and background color are often present in photomicrographs of the specimen (shown in Fig. 7.32(a)), and secondary fluorescence colors are degraded. Placing an auxiliary ultraviolet and/or short blue wavelength barrier filter somewhere in the microscope optical pathway is the method of choice for alleviating this problem.

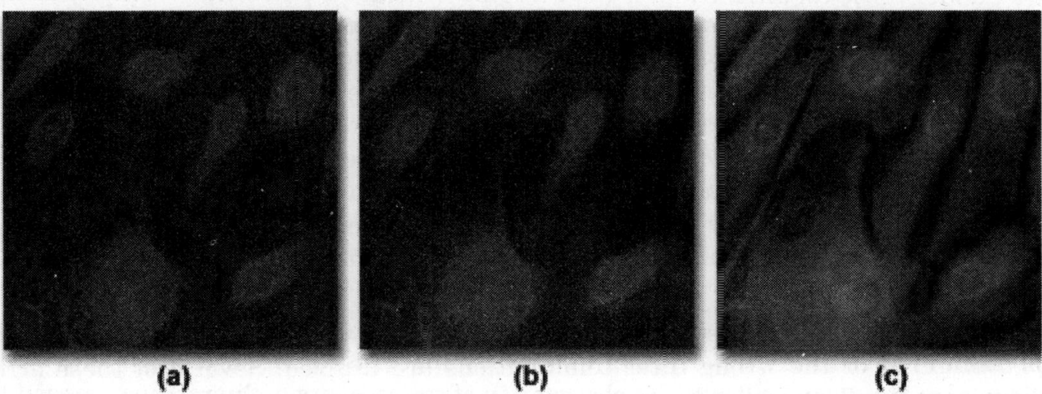

Fig. 7.32. Filter Errors in Fluorescence Photomicrography.

Fluorochromes used to stain specimens often produce an undesirable amount of short wavelength blue autofluorescence, particularly when staining tissue sections. As mentioned above, this autofluorescence must be absorbed by the barrier filter to avoid degradation of secondary fluorescence emission produced by the fluorochrome-tagged specimen. Kodak Wratten filters 2A and 2E, which absorb short-wavelength blue radiation, are suitable for use in this application. The microscope manufacturers and aftermarket suppliers also offer filters useful in fine-tuning the cutoff range of barrier filters.

Intrinsic autofluorescence in mounting media can also cause the type of problems associated with the photomicrograph illustrated in Fig. 7.32(a). Many of the popular permanent mounting media formulations exhibit autofluorescence as an undesirable side effect, and should not be used for quantitative fluorescence work. Instead, temporary mounts made with pure glycerin, glycerin-water mixtures, or fluorescence-free immersion oils should be utilized to prepare specimens. The microscope manufacturers offer specialized objectives that are optically corrected for specimen observation in glycerol. Autofluorescence produced by mounting media is usually pale blue or green in colour and will lead to degradation of secondary fluorescence by the specimen. If the barrier filter does not remove superfluous autofluorescence generated by mounting media, sometimes Kodak Wratten color compensating filters can be added to the pathway to alleviate this problem. The pale blue color generated by most media can be removed by using a filter such as the CC20Y, and pale green fluorescence can be neutralized with a CC20M filter. Use of higher density filters (CC30 and above) will often lead to absorption of secondary fluorescence.

Autofluorescence exhibited by cells and tissues in culture can often limit the ability to detect fluorescent probes in stained and fixed preparations. The extent of autofluorescence displayed by fixed tissues will often vary depending upon specimen preparation and the experimental conditions used to detect fluorescence. Flavin coenzymes (FAD and FMN) and reduced pyridine nucleotides (NADH) are the principal sources of autofluorescence in mammalian cells. Secondary fluorescence in fluorochrome-stained plant cells is often masked by autofluorescence from lignins (green autofluorescence) and porphyrins such as chlorophyll (red autofluorescence). This problem can sometimes be overcome with fixed cells and tissues by thorough washing with a 0.1 per cent solution of sodium borohydride in phosphate-buffered saline for 30 minutes prior to staining.

When excitation and/or emission filters are not matched to the wavelength absorption and emission spectrum of chromophores used to stain specimens, then secondary fluorescence is usually compromised, as illustrated in Fig. 7.32(c). In this example, an improper emission filter having too short of a wavelength cutoff is used to pass secondary fluorescence emitted by the specimen. Because a majority of this fluorescence is centered in the blue region of the spectrum (at 455 nanometers), only muddy reddish long wavelength color tones are passed by the filter. To correct the problem, a new barrier filter should be chosen. Overexposure of color film, along with reciprocity errors, can also cause dramatic color shifts of secondary fluorescence in final images.

Contrast errors often arise in fluorescence microscopy when the microscope optical train is mis-configured or the wrong filter combinations are utilized. Several of these errors are presented in Fig. 7.33, which illustrates poor contrast in secondary fluorescence images caused by improper excitation filter choice and fluorescent immersion oil. The specimen is a cross section of a Japanese maple leaf petiole undergoing autofluorescence and imaged with a 10x fluorite objective using an Olympus WBV filter cube. Under ideal conditions, the image presented in Fig. 7.33(b) would be obtained. However, when the excitation filter has too great a bandwidth (allowing unwanted wavelengths to pass), the image loses contrast resulting in the photomicrograph illustrated in Fig. 7.33(a). Choosing the correct filter combination requires knowledge of the spectral values of excitation and emission of the chromophore(s) used to stain the specimen. In this case, the excitation filter passes wavelengths up to 550 nanometers, interfering with green secondary fluorescence emitted by the specimen and severely compromising image contrast.

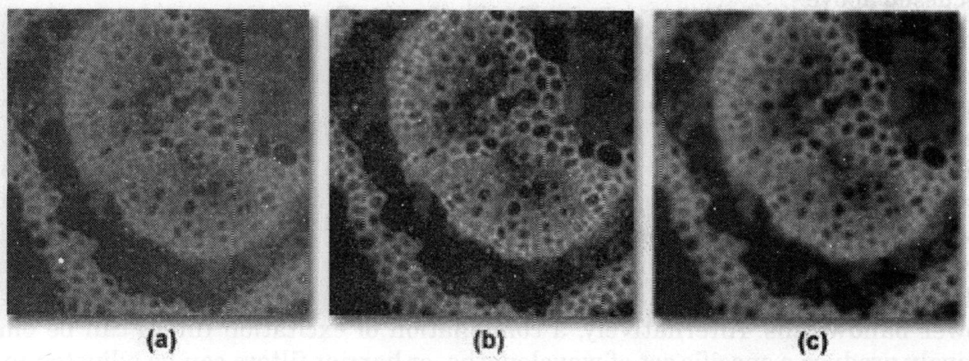

Fig. 7.33. Contrast Errors in Fluorescence Photomicrography.

Controlling specimen contrast in fluorescence microscopy requires that all optical components of the microscope be free of autofluorescence, and that they should not scatter light to a significant degree. This includes internal glass lens elements, mirrors, filters, and beamsplitters. Objectives should provide high image quality while transmitting light with high efficiency down into the near ultraviolet range. Filters must transmit desired wavelengths while blocking others that might interfere with observation of secondary fluorescence. Because excitation light is of far greater intensity than specimen fluorescence, the task of completely removing this light from the fluorescence image is not very easy and requires careful selection of filter combinations that provide a narrow range of excitation wavelengths (usually between 10 and 75 nanometers in bandwidth). When using high numerical aperture immersion objectives, it is critical that the immersion oil be free of contaminants that might exhibit fluorescence. Figure 7.33(c) illustrates an example of poor contrast generated by autofluorescence in immersion oil. Although the use of immersion oil helps to improve image brightness, partially by eliminating the loss of errant light caused by reflection from surfaces (in particular, the coverslip and glass lens elements), fluorescing contaminants that induce autofluorescence mix unwanted wavelengths with secondary fluorescence from the specimen. This has the unfortunate consequence of brightening the background while simultaneously reducing image contrast, both of which are undesirable.

Artifacts in specimen preparation will also produce contrast problems in fluorescence microscopy. Early pioneers in fluorescence often utilized non-specific fluorochromes that produced very bright images due to the fact that the dye molecules were randomly scattered throughout the specimen. Newer fluorescent techniques with improved and highly specific fluorochromes have resulted in specimens exhibiting stained target areas having substantially weaker fluorescence because much less dye is bonded. These newer methods require more careful specimen preparation to avoid unwanted fluorescence artifacts. It is often helpful to pre-stain portions of the specimen with non-fluorescing dyes that compete for binding sites to aid in fluorochrome selectivity. Specimens should be washed thoroughly to remove unbound fluorochrome before placement on acid-cleaned microscope slides made with non-fluorescent glass. Optimization of the fluorochrome environment by addition of chemical additives to minimize photobleaching helps to reduce or eliminate specimen fading. In addition, careful selection of the mounting medium chemical composition and pH is important to obtain the best image from secondary fluorescence. Coverslips and immersion oil should also be non-fluorescent, as discussed above.

Autofluorescence and/or non-specific fluorescence can occur as a result of the inherent properties of the specimen or through errors in staining. Typical errors of this type are illustrated in Figure 7.34(a), which shows a thin section of rat colon tissue stained with a mixture of fluorescein-5-isothiocyanate (FITC) and 4',6-diamidino-2-phenylindole (DAPI). The overall blue cast in the photomicrograph is due to a combination of autofluorescence and non-specific fluorescence due to overstaining with the DAPI fluorochrome. Proper staining with both dyes, followed by thorough washing of the specimen results in the photomicrograph presented in Figure 7.34(b). Errors similar to that illustrated in Figure 7.34(a) are also caused by using too broad an excitation bandwidth, a reversible mistake that can be corrected by use of filters with narrower bandwidths. Alternatively, a combination of excitation filters can be employed to selectively produce a specific set of wavelengths, or barrier filters can be adjusted to pass only longer wavelengths. Figure 8(c) illustrates the rat colon thin section when stained properly but now with a light red background caused by lack of a red suppression filter in the optical pathway. This error can be corrected by placing the proper suppression filter in the optical path.

Fluorescence photomicrography also suffers from common problems encountered with other forms of optical microscopy and photomicrography. These include focus errors caused by improper adjustment of the focal distance between the optics and film plane, microscope stand vibration, contamination by dirt and debris, underexposure, overexposure, color balance errors, specular reflections (common with long exposures and overhead room lighting), vignetting, and optical microscope mis-alignment. Examples of these errors are presented in our section entitled.

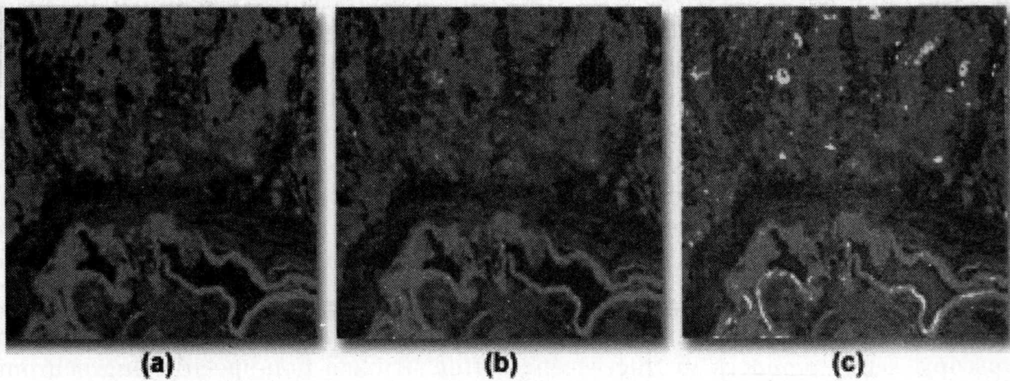

Fig. 7.34. Autofluoresncence and Non-Specific Fluorescence Errors in Fluorescence Photomicrography.

Troubleshooting Photomicrography Errors

Cameras and Film for Fluorescence Photomicrography : Improvements in film technology have produced increased film speeds and resolution improvements, but there are no color films currently available uniquely suited to fluorescence photomicrography. The specific properties of individual color films, including the characteristic curves for individual emulsion layers, properties of coupled dyes, and the spectral response to light used for fluorescence excitation and emission often vary dramatically from film to film and are sometimes difficult to control and reproduce. These problems lead to difficulty in maintaining accurate color rendering and exposure with color photomicrography using film.

Exposure measurement in fluorescence photomicrography is complicated by the fact that the brightness range within a fluorescent specimen is usually much greater than can be accurately recorded on film. Overexposure of the highlights will result in loss of fine detail and colour saturation, while underexposure produces very dark images that do not reveal features hidden in shadowed areas. The background, which is usually very dark or black, is not as critical to either over- or under-exposure. Three primary methods are utilized to determine exposure when using fluorescence: brightness measurement with a CCD, photodiode, or photomultiplier, continuous measurement during exposure (photomultiplier), and the least desirable method, trial and error. Modern photomicrography systems designed for fluorescence are equipped with the ability to gauge illumination intensity and calculate exposure times based on this reading. For routine fluorescence photomicrography, some cameras have a silicon blue detector cell for determining exposure duration, while others have a photomultiplier metering detector for measuring extremely faint specimens.

Table 7.5. Film Properties.

ISO Number	*Light Sensitivity*	*Resolution*	*Grain Size*	*Exposure Tolerance*
25-50	Low	High	Fine	Wide
100-200	Medium	Medium	Medium	Medium
400+	High	Low	Coarse	Low

It is important to consider several factors in choosing a detector system for fluorescence photomicrography. These include the detector's noise level and the signal-to-noise ratio for the image, spatial and temporal resolution, geometric distortion, linearity, and spectral sensitivity. In determining proper exposure for fluorescing specimens, it is best to use a camera with a CCD or photomultiplier metering system capable of accurate low-light measurement. Spot metering capability is valuable for measuring the area actually fluorescing and not the unwanted metering of the usual dark background surround the fluorescing material. Another technique involves the use of the exposure adjustment control to reduce exposure times that would otherwise be unduly lengthened by measuring the dark background of the field of view. In this respect, Olympus and Nikon offer camera systems that have built-in algorithms for reducing exposure if the camera is set for fluorescence mode. When light is scant (usually most of the time with fluorescence), set the trinocular prism to send all the light up to the film plane to reduce exposure time.

Ideally, a spot meter should be included in the photomicrography system to aid in determining exposure when only a few bright objects are visible on an otherwise dark background. For integrated light measurements made over the entire viewfield without a spot meter, the exposure time should be cut in half or to a fourth of the metered reading (for scattered fluorescing objects on a dark background), much like the situation with darkfield microscopy. If the spot meter light sensor is in a fixed position in the middle of the viewfield, the specimen must first be moved to the center of the field to obtain a suitable area for exposure measurement, then relocated to compose the photomicrograph. High-end camera systems allow translation of the spot meter sensor to enable the microscopist to determine photomicrograph composition prior to exposure measurement. After positioning of the specimen for photomicrography, the spot meter sensor can then be moved to the brightest area in the viewfield for exposure measurement without disturbing the specimen. When using a spot meter, make certain the fluorescent object is large enough to fill the spot and record an accurate exposure reading.

Some photomicrography camera systems provide a permanent beamsplitter assembly that directs a portion of the light to a photometer, which allows light measurement during the actual exposure in order to compensate for fading. This is a convenient method for adjusting exposure times on the fly to avoid photobleaching artifacts, but it has the disadvantage of reducing light intensity directed to the film, necessitating a longer exposure. More advanced computer-aided camera systems are able to compensate for reciprocity failure and fading through a pre-programmed lookup table conveniently stored in the metering system's memory. Automatic correction for reciprocity failure by camera system computers operate by applying the following equation to exposure calculations:

$$\mathbf{T}_c = (\mathbf{T}_m)^p$$

where $\mathbf{T}_c$ is the corrected exposure time, $\mathbf{T}_m$ is the metered (uncorrected) time, and $\boldsymbol{p}$ is a constant determined from reciprocity values for individual films housed in the system's lookup table. This system is limited by the amount and type of reciprocity data for films that are added to the lookup table during initial programming of the metering system.

Without an integral exposure metering system, determination of exposure time must be made by trail and error. This method involves making a series of trial exposures at a variety of exposure times, stepped through increments of one-half to a full f-stop (bracketing). After processing the film, the exposure that produces the best result is carefully noted and provides the basis for further photomicrography using the identical microscope configuration. Changes to specimen illumination or microscope setup will necessitate another series of exposure brackets to determine the new optimum exposure time. If only the objective magnification is changed, new exposure times can be calculated from previously recorded bracket data. Many photomicrographers doing fluorescence microscopy bracket their exposures, both over and under by three or more intervals, to make certain that they will secure a correct exposure at one of the settings. Keeping good written records of exposure and associated equipment data can reduce or eliminate the need for future bracketing. When determining exposure using the trial and error method, it is wise to continuously monitor the microscope lamp output for deviations in illumination that occur as the lamp ages. Microscopists using this method of exposure determination often rely on SLR cameras to capture images with fluorescence illumination. With these cameras, movement of the shutter and viewfinder mirror during exposure may cause vibrations that lead to unsharp images. This problem can usually be circumvented by using a remote shutter switch and/or raising the mirror prior making the exposure. In some cameras, operation of the self-timer releases the mirror, which allows time for vibrations to cease before the exposure is made.

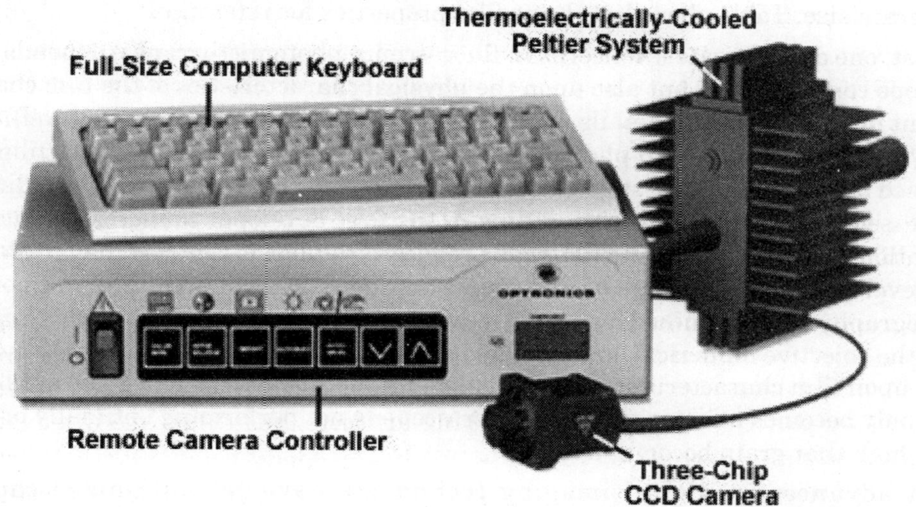

Fig. 7.35. Digital Camera for Fluorescence Photomicrography.

Microscopes equipped specifically for fluorescence illumination often have an illuminated reticle in the system or focusing telescope, which allows the microscopist to visualize a photomicrography grating superimposed over faintly visible specimens positioned against a very dark background. Both Olympus and Nikon provide reticles in their high-end microscopes that offer a choice of crosshair and film frame colors with which to view the specimen. Microscopists can choose between a red or yellow illuminated reticle. The red reticle has the advantage of maintaining the dark adaptation of the observer's eye and provides excellent contrast when viewed over blue and green secondary fluorescence. For red fluorescence and when simultaneous phase contrast or DIC are used, the yellow reticle should be selected to maximize contrast. Some manufacturers offer a focusing telescope with illuminated framing reticle that attaches to the automatic camera housing and allows viewing of the film frame dimensions over the dark background field.When choosing film for fluorescence photomicrography, the film speed (ASA or ISO rating) is a variable that should be carefully considered. The higher the ISO rating, other things being equal, the less the amount of light necessary to register a satisfactory image on the film. Illumination sources used in fluorescence microscopy require daylight-balanced film emulsions to accurately render specimen secondary fluorescence colors. Most major film manufacturers offer a wide spectrum of daylight films at a variety of ISO ratings in both transparency and color negative formats. Kodak daylight Ektachrome with an ISO of 200 or 400 or Kodachrome 200, both in 35 millimeter size, are positive transparency films that display adequate speed, good color rendition and acceptable resolution. Fujichrome Provia is offered in 400 and 1600 ISO ratings, and is an excellent transparency film in terms of color saturation and resolution. Other manufacturers make comparable transparency films, but we suggest avoiding the use of color negative film if at all possible. For black and white film, Kodak offers T-Max 400, an excellent fine to medium grain film that is one of the best for fluorescence applications. Several black & white films are available that can be processed with standard color negative chemicals (C41). These include Kodak T-Max 400CN, Kodak Advantix, and Ilford XP-2 Super, all of which are offered in ISO ratings of 400. Keep in mind that faster films allow for much shorter exposure times at the expense of increased grain size. Table above lists basic film properties for reference.

Contrast, one of the greatest concerns in fluorescence photomicrography, depends not only on microscope configuration, but also upon the physical characteristics of the film chosen, the development process, exposure details, and the specimen itself with respect to the efficiency of staining by the fluorochrome and photobleaching or fading. Colour transparency films can be underexposed by one or several f-stops (a bonus when considering reciprocity and fading), then **push** processed by extending processing time in the first developer to increase contrast and color saturation. Many black & white films can produce variations in contrast that are dependent upon the developer mixing conditions and processing time. Resolution of film in fluorescence photomicrography is determined principally by the optical parameters of the microscope (primarily the objective numerical aperture) and how well it is configured, but is also somewhat dependent upon film characteristics and development methods. In practice however, the film resolution only becomes a factor when the microscope is not performing optimally or the ISO speed is so high that grain becomes a factor.

Recent advances in digital imaging technology have revolutionized capture of photomicrographs using charge-coupled devices (CCDs). When imaging multiply-stained fluorescence specimens, digital cameras provide an excellent solution and are rapidly becoming the medium of choice for many photomicrographers. The price of three-chip integrating color CCD cameras has dropped to an affordable range and, when coupled to a frame-grabber or digital video card and computer, offer an excellent method of collecting, digitizing, editing, and storing fluorescence images.

The weak fluorescence displayed by many specimens necessitates high performance in CCD detector specifications. Currently, the best choice is a back-thinned frame or interline transfer device, which is equipped with heat sinks and/or external reservoirs for cooling. Many of these cameras have a quantum efficiency approaching or exceeding 90 per cent and permit the sequential acquisition of several images before readout. Image acquisition cycles run from one to five seconds, depending on the readout rate, video capture parameters, and CCD resolution. A typical CCD camera matching these specifications is the **Optronics** MagnaFire digital imaging camera system designed for scientific applications. This camera has a SONY ICX085AL 2/3-inch interline transfer CCD with an array size of 1300 × 1030 and a (square) pixel size of 6.7 microns. Camera performance is enhanced with a 60 dB dynamic range and signal-to-noise ratio, a file bit depth of 24, 32, or 48 bit color, a dark current of 4 electrons per pixel/ second with a 16,000 electron well depth, and sealed gas Peltier CCD cooling. Color image formation is via multiple dichroic optical glass color filters. Optronics and other manufacturers also offer a wide spectrum of digital cameras (a typical CCD camera is illustrated in Figure 84) suited to almost every need in both fluorescence and routine optical microscopy.

The following points summarize important aspects of fluorescence photomicrography.

- Use high numerical aperture objectives equipped with glass or quartz lenses that are transparent to near ultraviolet light. Also use the lowest magnification projection lens possible to limit the total magnification.

- Reflected light fluorescence is preferred to transmitted light fluorescence. If it is not possible to configure the microscope for reflected light, use a transmitted light dark field condenser.
- Optimize specimen contrast by careful choice of film parameters, proper microscope configuration, and by carefully matching chromophore excitation and secondary fluorescence properties to the proper excitation and barrier filters.
- Remove the trinocular beam splitter prism from the light path to allow 100 per cent of the light to reach the film plane during exposure. Also use 35 millimeter film and avoid larger film formats.
- Choose daylight balanced transparency films for the best rendition of specimen colors, resolution, and contrast. Push process transparency films for enhanced contrast and color saturation.
- The light source must provide high-intensity radiation in very narrow ranges of the spectrum, typically between 10 and 50 nanometers. The best choices are xenon and mercury vapor lamps or lasers tuned to specific wavelengths.
- The microscope optical train, including lenses, mirrors, filters, beam splitters, immersion media, microscope slides, and cover slips, should be free of auto fluorescing components.
- When not viewing or photographing fluorescence specimens, block the excitation light using the filter slider or shutter in the illuminator to avoid specimen damage by photo bleaching.
- Always provide a heat filter between the illuminator and fluorescence filters, which may be damaged by excess heat.
- Monitor the lamp for signs of flicker and replace when illumination intensity becomes uneven. Keep the lamp and vertical illuminator adjusted for proper Köhler illumination.
- Prepare specimens carefully and remove excess chromophores by thorough washing after staining.

In conclusion, the field of fluorescence microscopy is rapidly expanding in many medical and biological research laboratories. The mainstream of fluorescence microscopy has undergone an almost total shift from utilizing transmitted light to incident light, accompanied by the introduction of many new and different fluorochromes. This has been an impetus for spurring many new and varied fluorescence microscopy applications.

CONFOCAL MICROSCOPY

It is an optical imaging technique used to increase micrograph contrast and/or to reconstruct three-dimensional images by using a spatial pinhole to eliminate out-of-focus light or flare in specimens that are thicker than the focal plane. This technique has gained popularity in the scientific and industrial communities. Typical applications include life sciences and semiconductor inspection.

Basic Concept

The principle of confocal imaging was patented by Marvin Minsky in 1957. In a conventional (i.e., wide-field) fluorescence microscope, the entire specimen is flooded in light from a light source. Due to the conservation of light intensity transportation, all parts of the specimen throughout the optical path will be excited and the fluorescence detected by a photodetector or a camera. In contrast, a confocal microscope uses point illumination and a pinhole in an optically conjugate plane in front of the detector to eliminate out-of-focus information. Only the light within the focal plane can be detected, so the image quality is much better than that of wide-field images. As only one point is illuminated at a time in confocal microscopy, 2D or 3D imaging requires scanning over a regular raster (i.e. a rectangular pattern of parallel scanning lines) in the specimen. The thickness of the focal plane is defined mostly by the inverse of the square of the numerical aperture of the objective lens, and also by the optical properties of the specimen and the ambient index of refraction. These microscopes also are able to see into the image by taking images at different depths. Fortuitously, shortly after Minsky's patent had expired, practical laser scanning confocal microscope designs were translated into working instruments by several investigators. Dutch physicist G. Fred Brakenhoff developed a scanning confocal microscope in 1979, while almost simultaneously, Colin Sheppard contributed to the technique with a theory of image formation. Tony Wilson, Brad Amos, and John White nurtured the concept and later (during the late 1980s) demonstrated the utility of confocal imaging in the examination of fluorescent biological specimens. The first commercial instruments appeared in 1987. During the 1990s, advances in optics and electronics afforded more stable and powerful lasers, high-efficiency scanning mirror units, high-throughput fiber optics, better thin film dielectric coatings, and detectors having reduced noise characteristics. In addition, fluorochromes that were more carefully matched to laser excitation lines were beginning to be synthesized. Coupled to the rapidly advancing computer processing speeds, enhanced displays, and large-volume storage technology emerging in the late 1990s, the stage was set for a virtual explosion in the number of applications that could be targeted with laser scanning confocal microscopy.

Modern confocal microscopes can be considered as completely integrated electronic systems where the optical microscope plays a central role in a configuration that consists of one or more electronic detectors, a computer (for image display, processing, output, and storage), and several laser systems combined with wavelength selection devices and a beam scanning assembly. In most cases, integration between the various components is so thorough that the entire confocal microscope is often collectively referred to as a digital or video imaging system capable of producing electronic images. These microscopes are now being employed for routine investigations on molecules, cells, and living tissues that were not possible just a few years ago.

Types

Three types of confocal microscopes are commercially available: Confocal laser scanning microscopes, spinning-disk (Nipkow disk) confocal microscopes and Programmable Array Microscopes (PAM). Confocal laser scanning microscopy yields better image quality than Nipkow and PAM, but the imaging frame rate was very slow (less than 3 frames/second) until recently; spinning-disk confocal microscopes can achieve video rate imaging—a desirable feature for dynamic observations such as live cell imaging. Confocal laser scanning microscopy has now been improved to provide better than video rate (60 frames/second) imaging by using MEMS based scanning mirrors.

The confocal principle in epi-fluorescence laser scanning microscopy is diagrammatically presented in Fig. 7.30. Coherent light emitted by the laser system (excitation source) passes through a pinhole aperture that is situated in a conjugate plane (confocal) with a scanning point on the specimen and a second pinhole aperture positioned in front of the detector (a photomultiplier tube). As the laser is reflected by a dichromatic mirror and scanned across the specimen in a defined focal plane, secondary fluorescence emitted from points on the specimen (in the same focal plane) pass back through the dichromatic mirror and are focused as a confocal point at the detector pinhole aperture.

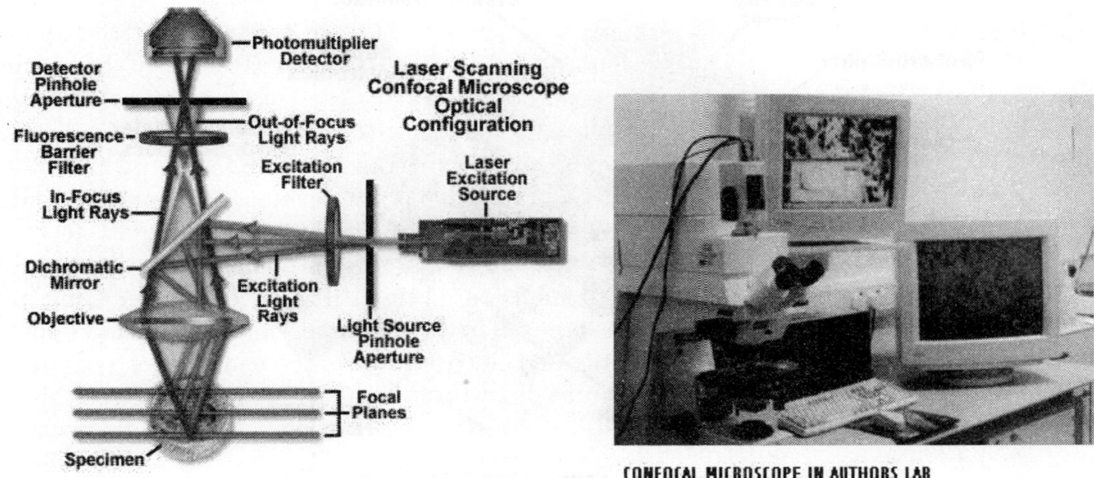

Fig. 7.36. The Principle of Confocal Microscopy.

The significant amount of fluorescence emission that occurs at points above and below the objective focal plane is not confocal with the pinhole (termed **Out-of-Focus Light Rays** in Fig. 7.30) and forms extended Airy disks in the aperture plane. Because only a small fraction of the out-of-focus fluorescence emission is delivered through the pinhole aperture, most of this extraneous light is not detected by the photomultiplier and does not contribute to the resulting image. The dichromatic mirror, barrier filter, and excitation filter perform similar functions to identical components in a widefield epi-fluorescence microscope. Refocusing the objective in a confocal microscope shifts the excitation and emission points on a specimen to a new plane that becomes confocal with the pinhole apertures of the light source and detector.

In traditional widefield epi-fluorescence microscopy, the entire specimen is subjected to intense illumination from an incoherent mercury or xenon arc-discharge lamp, and the resulting image of secondary fluorescence emission can be viewed directly in the eyepieces or projected onto the surface of an electronic array detector or traditional film plane. In contrast to this simple concept, the mechanism of image formation in a confocal microscope is fundamentally different. As discussed above, the confocal fluorescence microscope consists of multiple laser excitation sources, a scan head with optical and electronic components, electronic detectors (usually photomultipliers), and a computer for acquisition, processing, analysis, and display of images.

The scan head is at the heart of the confocal system and is responsible for rasterizing the excitation scans, as well as collecting the photon signals from the specimen that are required to assemble the final image. A typical scan head contains inputs from the external laser sources, fluorescence filter sets and dichromatic mirrors, a galvanometer-based raster scanning mirror system, variable pinhole apertures for generating the confocal image, and photomultiplier tube detectors tuned for different fluorescence wavelengths. The general arrangement of scan head components is presented in Figure 86 for a typical commercial unit.

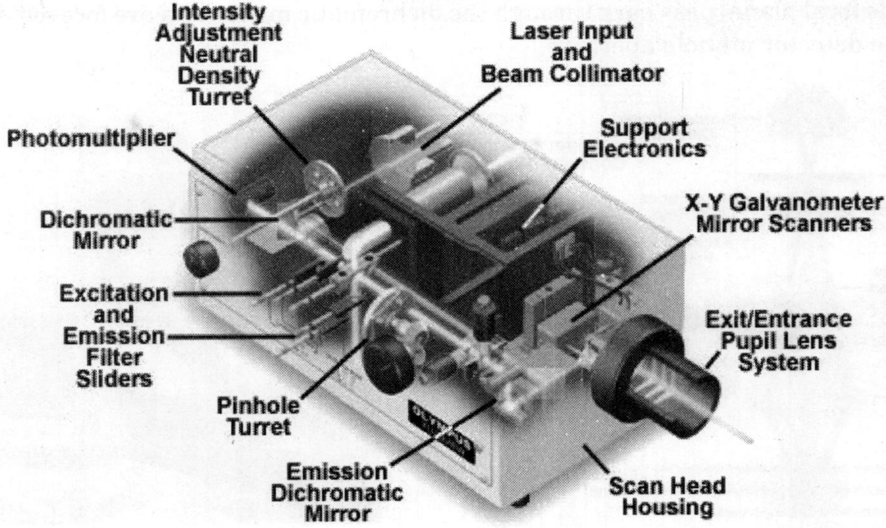

Fig. 7.37. Confocal Microscopy Scanning unit.

In epi-illumination scanning confocal microscopy, the laser light source and photomultiplier detectors are both separated from the specimen by the objective, which functions as a well-corrected condenser and objective combination. Internal fluorescence filter components (such as the excitation and barrier filters and the dichromatic mirrors) and neutral density filters are contained within the scanning unit (see Figure 7.37). Interference and neutral density filters are housed in rotating turrets or sliders that can be inserted into the light path by the operator. The excitation laser beam is connected to the scan unit with a fiber optic coupler followed by a beam expander that enables the thin laser beam wrist to completely fill the objective rear aperture (a critical requirement in confocal microscopy). Expanded laser light that passes through the microscope objective forms an intense diffraction-limited spot that is scanned by the coupled galvanometer mirrors in a raster pattern across the specimen plane (point scanning).

One of the most important components of the scanning unit is the pinhole aperture, which acts as a spatial filter at the conjugate image plane positioned directly in front of the photomultiplier. Several apertures of varying diameter are usually contained on a rotating turret that enables the operator to adjust pinhole size (and optical section thickness). Secondary fluorescence collected by the objective is **descanned** by the same galvanometer mirrors that form the raster pattern, and then passes through a barrier filter before reaching the pinhole aperture. The aperture serves to exclude fluorescence signals from out-of-focus features positioned above and below the focal plane, which are instead projected onto the aperture as Airy disks having a diameter much larger than those forming the image. These oversized disks are spread over a comparatively large area so that only a small fraction of light originating in planes away from the focal point passes through the aperture. The pinhole aperture also serves to eliminate much of the stray light passing through the optical system. Coupling of aperture-limited point scanning to a pinhole spatial filter at the conjugate image plane is an essential feature of the confocal microscope.

When contrasting the similarities and differences between widefield and confocal microscopes, it is often useful to compare the character and geometry of specimen illumination utilized for each of the techniques. Traditional widefield epi-fluorescence microscope objectives focus a wide cone of illumination over a large volume of the specimen, which is uniformly and simultaneously illuminated. A majority of the fluorescence emission directed back towards the microscope is gathered by the objective (depending upon the numerical aperture) and projected into the eyepieces or detector. The result is a significant amount of signal due to emitted background light and autofluorescence originating from areas above and below the focal plane, which seriously reduces resolution and image contrast.

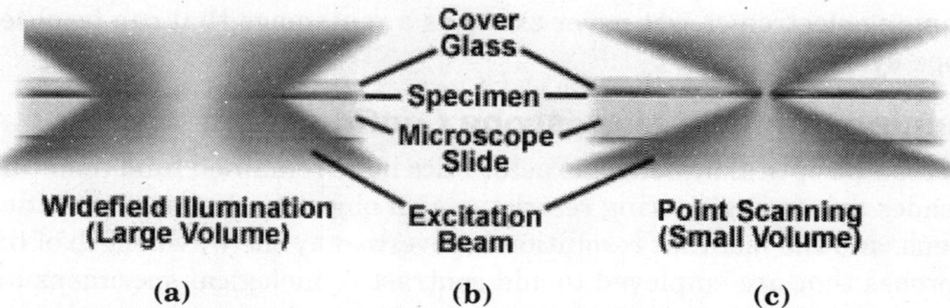

Fig. 7.38. Widefield versus Confocal Point Scanning of Specimens.

The laser illumination source in confocal microscopy is first expanded to fill the objective rear aperture, and then focused by the lens system to a very small spot at the focal plane (Fig. 7.38(b)). The size of the illumination point ranges from approximately 0.25 to 0.8 micrometers in diameter (depending upon the objective numerical aperture) and 0.5 to 1.5 micrometers deep at the brightest intensity. Confocal spot size is determined by the microscope design, wavelength of incident laser light, objective characteristics, scanning unit settings, and the specimen. Presented in Figure 4 is a comparison between the typical illumination cones of a widefield (Figure 7.38(a)) and point scanning confocal (Fig. 7.38(b)) microscope at the same numerical aperture. The entire depth of the specimen over a wide area is illuminated by the widefield microscope, while the sample is scanned with a finely focused spot of illumination that is centered in the focal plane in the confocal microscope.

In laser scanning confocal microscopy, the image of an extended specimen is generated by scanning the focused beam across a defined area in a raster pattern controlled by two high-speed oscillating mirrors driven by galvanometer motors. One of the mirrors moves the beam from left to right along the **x** lateral axis, while the other translates the beam in the **y** direction. After each single scan along the **x** axis, the beam is rapidly transported back to the starting point and shifted along the **y** axis to begin a new scan in a process termed **flyback**. During the flyback operation, image information is not collected. In this manner, the area of interest on the specimen in a single focal plane is excited by laser illumination from the scanning unit.

As each scan line passes along the specimen in the lateral focal plane, fluorescence emission is collected by the objective and passed back through the confocal optical system. The speed of the scanning mirrors is very slow relative to the speed of light, so the secondary emission follows a light path along the optical axis that is identical to the original excitation beam. Return of fluorescence emission through the galvanometer mirror system is referred to as

descanning. After leaving the scanning mirrors, the fluorescence emission passes directly through the dichromatic mirror and is focused at the detector pinhole aperture. Unlike the raster scanning pattern of excitation light passing over the specimen, fluorescence emission remains in a steady position at the pinhole aperture, but fluctuates with respect to intensity over time as the illumination spot traverses the specimen producing variations in excitation.

Fluorescence emission that is passed through the pinhole aperture is converted into an analog electrical signal having a continuously varying voltage (corresponding to intensity) by the photomultiplier. The analog signal is periodically sampled and converted into pixels by an analog-to-digital (**A/D**) converter housed in the scanning unit or the accompanying electronics cabinet. The image information is temporarily stored in an image frame buffer card in the computer and displayed on the monitor. It is important to note that the confocal image of a specimen is reconstructed, point by point, from emission photon signals by the photomultiplier and accompanying electronics, yet never exists as a real image that can be observed through the microscope eyepieces.

Laser Scanning Confocal Microscope Configuration

Basic microscope optical system characteristics have remained fundamentally unchanged for many decades due to engineering restrictions on objective design, the static properties of most specimens, and the fact that resolution is governed by the wavelength of light. However, fluorescent probes that are employed to add contrast to biological specimens and, and other technologies associated with optical microscopy techniques, have improved significantly. The explosive growth and development of the confocal approach is a direct result of a renaissance in optical microscopy that has been largely fueled by advances in modern optical and electronics technology. Among these are stable multi-wavelength laser systems that provide better coverage of the ultraviolet, visible, and near-infrared spectral regions, improved interference filters (including dichromatic mirrors, barrier, and excitation filters), sensitive low-noise wide band detectors, and far more powerful computers. The latter are now available with relatively low-cost memory arrays, image analysis software packages, high-resolution video displays, and high quality digital image printers. The flow of information through a modern confocal microscope is presented diagrammatically in Fig. 7.39.

Although many of these technologies have been developed independently for a variety of specifically-targeted applications, they have been gradually been incorporated into mainstream commercial confocal microscopy systems. In current microscope systems, classification of designs is based on the technology utilized to scan specimens. Scanning can be accomplished either by translating the stage in the **x**, **y**, and **z** directions while the laser illumination spot is held in a fixed position, or the beam itself can be raster-scanned across the specimen. Because three-dimensional translation of the stage is cumbersome and prone to vibration, most modern instruments employ some type of beam-scanning mechanism.

In modern confocal microscopes, two fundamentally different techniques for beam scanning have been developed. **Single-beam** scanning, one of the more popular methods employed in a majority of the commercial laser scanning microscopes, uses a pair of computer-controlled galvanometer mirrors to scan the specimen in a raster pattern at a rate of approximately one frame per second. Faster scanning rates (to near video speed) can be achieved using acousto-optic devices or oscillating mirrors. In contrast, **multiple-beam** scanning confocal microscopes are equipped with a spinning **Nipkow** disk containing an array of pinholes and microlenses. These instruments often use arc-discharge lamps for illumination instead of lasers to reduce specimen damage and enhance the detection of low fluorescence levels during real time image collection. Another important feature of the multiple-beam microscopes is their ability to readily capture images with an array detector, such as a charge-coupled device (**CCD**) camera system.

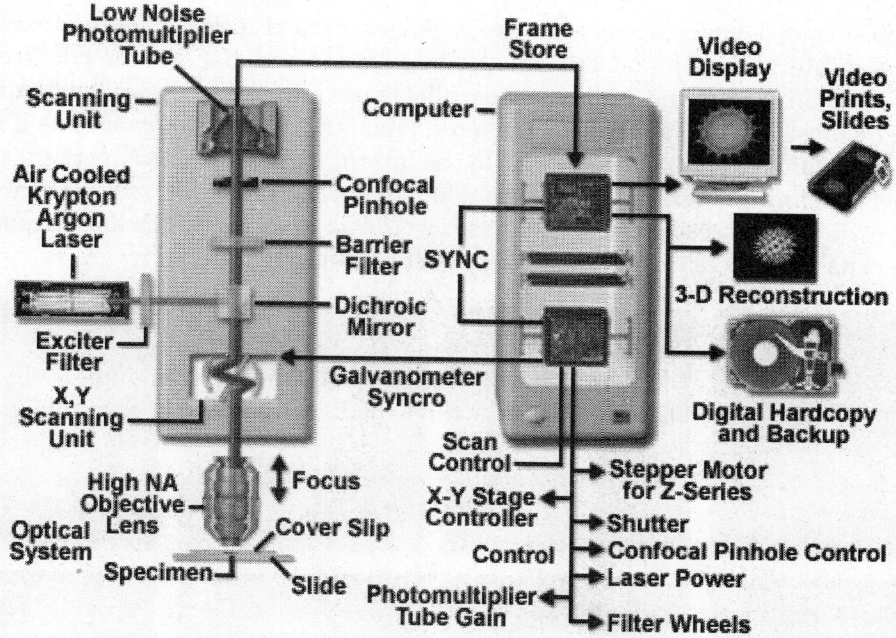

Fig. 7.39. Confocal Microscopy Information Flows Schemeatic Diagram.

All laser scanning confocal microscope designs are centered around a conventional upright or inverted research-level optical microscope. However, instead of the standard tungsten-halogen or mercury arc-discharge lamp, one or more laser systems are used as a light source to excite fluorophores in the specimen. Image information is gathered point by point with a specialized detector such as a photomultiplier tube or avalanche photodiode, and then digitized for processing by the host computer, which also controls the scanning mirrors and/or other devices to facilitate the collection and display of images. After a series of images (usually serial optical sections) has been acquired and stored on digital media, analysis can be conducted utilizing numerous image processing software packages available on the host or a secondary computer.

Advantages and Disadvantages of Confocal Microscopy

The primary advantage of laser scanning confocal microscopy is the ability to serially produce thin (0.5 to 1.5 micrometer) optical sections through fluorescent specimens that have a thickness ranging up to 50 micrometers or more. The image series is collected by coordinating incremental changes in the microscope fine focus mechanism (using a stepper motor) with sequential image acquisition at each step. Image information is restricted to a well-defined plane, rather than being complicated by signals arising from remote locations in the specimen. Contrast and definition are dramatically improved over widefield techniques due to the reduction in background fluorescence and improved signal-to-noise. Furthermore, optical sectioning eliminates artifacts that occur during physical sectioning and fluorescent staining of tissue specimens for traditional forms of microscopy. The non-invasive confocal optical sectioning technique enables the examination of both living and fixed specimens under a variety of conditions with enhanced clarity.

With most confocal microscopy software packages, optical sections are not restricted to the perpendicular lateral (**x-y**) plane, but can also be collected and displayed in transverse planes. Vertical sections in the **x-z** and **y-z** planes (parallel to the microscope optical axis) can be readily generated by most confocal software programs. Thus, the specimen appears as if it had been sectioned in a plane that is perpendicular to the lateral axis. In practice, vertical sections are obtained by combining a series of **x-y** scans taken along the **z** axis with the software, and then projecting a view of fluorescence intensity as it would appear should the microscope hardware have been capable of physically performing a vertical section.

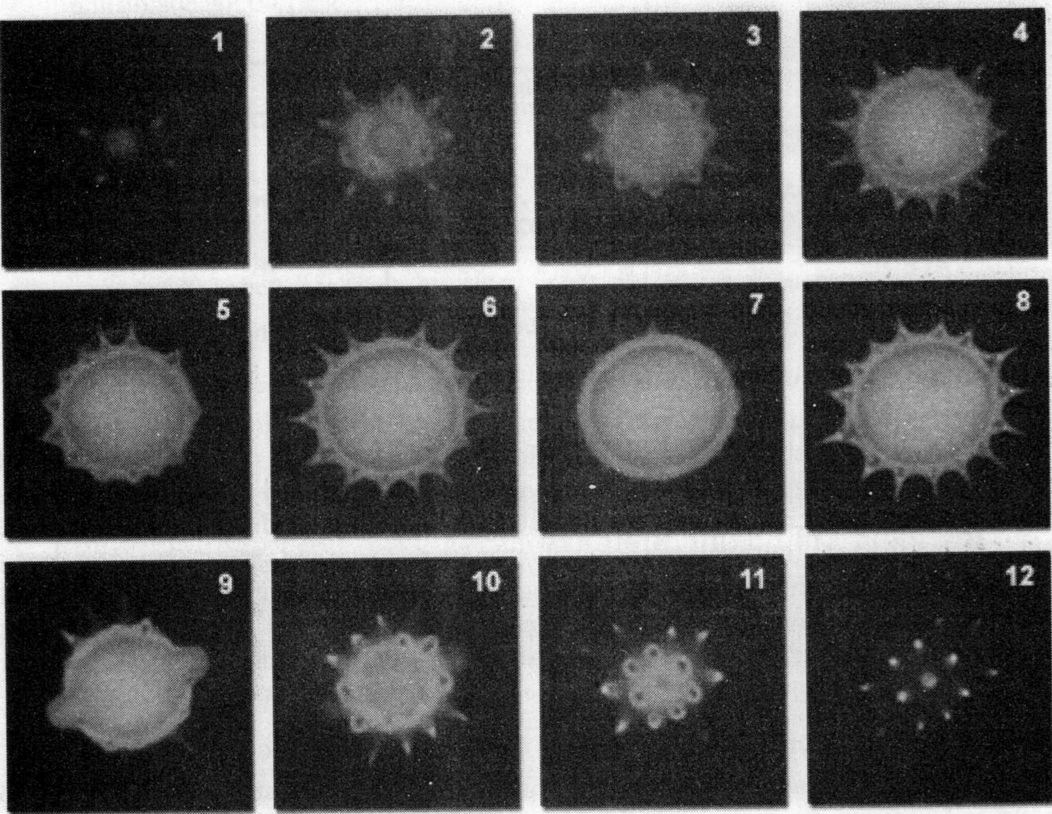

Fig. 7.40. Pollen Grain Serial Optical Sections by Confocal Microscopy.

A typical stack of optical sections (often termed a **z-series**) through a sunflower pollen grain revealing internal variations in autofluorescence emission wavelengths is illustrated in Fig. 7.40. Optical sections were gathered in 0.5-micrometer steps perpendicular to the z-axis (microscope optical axis) using a dual argon-ion (488 nanometer; green fluorescence) and green helium/neon (543 nanometer; red fluorescence) laser system. Pollen grains of from this species range between 20 and 40 micrometers in diameter and yield blurred images in widefield fluorescence microscopy, which lack information about internal structural details. Although only 12 of the over 48 images collected through this series are presented in the figure, they represent individual focal planes separated by a distance of approximately 3 micrometers and provide a good indication of the internal grain structure.

In specimens more complex than a pollen grain, complex interconnected structural elements can be difficult to discern from a large series of optical sections sequentially acquired through the volume of a specimen with a laser scanning confocal microscope. However, once an adequate series of optical sections has been gathered, it can be further processed into a three-dimensional representation of the specimen using volume-rendering computational techniques. This approach is now in common use to help elucidate the numerous interrelationships between structure and function of cells and tissues in biological investigations. In order to ensure that adequate data is collected to produce a representative volume image, the optical sections should be recorded at the appropriate axial (**z**-step) intervals so that the actual depth of the specimen is reflected in the image.

Most of the software packages accompanying commercial confocal instruments are capable of generating composite and multi-dimensional views of optical section data acquired from z-series image stacks. The three-dimensional software packages can be employed to create either a single three-dimensional representation of the specimen (Figure 7.41 or a video (movie) sequence compiled from different views of the specimen volume. These sequences often mimic the effect of rotation or similar spatial transformation that enhances the appreciation of the specimen's three-dimensional character. In addition, many software packages enable investigators to conduct measurements of length, volume, and depth, and specific parameters of the images, such as opacity, can be interactively altered to reveal internal structures of interest at differing levels within the specimen.

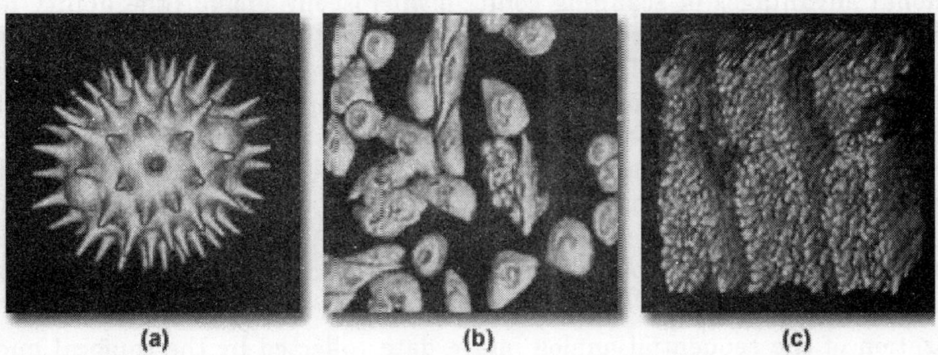

Fig. 7.41. Three-Dimensional Volume Renders from Confocal Optical Sections.

Typical three-dimensional representations of several specimens examined by serial optical sectioning are presented in Figure 90. The algorithm utilized to construct the three-dimensional model enables the user to rotate the pollen grain through 360 degrees for examination. The tissue culture cells in Fig. 7.41(b) are derived from the Chinese hamster ovary (**CHO**) line and were transfected with a chimeric plasmid vector containing the green fluorescent protein and a human immunodeficiency virus (**HIV**) protein that is expressed in the nucleus (thus, labelling the nuclear region). Thick tissue sections are also easily viewed in three-dimensions constructed from optical sections. The mouse intestine section illustrated in Figure 7.41 (c) was labeled with several fluorophores and created from a stack of 45 optical sections.

In many cases, a **composite** or **projection view** produced from a series of optical sections provides important information about a three-dimensional specimen than a multi-dimensional view. For example, a fluorescently labeled neuron having numerous thin, extended processes in a tissue section is difficult (if not impossible) to image using widefield techniques due to out-of-focus blur. Confocal thin sections of the same neuron each reveal portions of several extensions, but these usually appear as fragmented streaks and dots and lack continuity. Composite views created by flattening a series of optical sections from the neuron will reveal all of the extended processes in sharp focus with well-defined continuity. Structural and functional analysis of other cell and tissue sections also benefits from composite views as opposed to, or coupled with, three-dimensional volume rendering techniques.

Advances in confocal microscopy have made possible **multi-dimensional** views of living cells and tissues that include image information in the **x**, **y**, and **z** dimensions as a function of time and presented in multiple colors (using two or more fluorophores). After volume processing of individual image stacks, the resulting data can be displayed as three-dimensional multicolor video sequences in real time. Note that unlike conventional widefield microscopy, all fluorochromes in multiply labelled specimens appear in register using the confocal microscope. Temporal data can be collected either from time-lapse experiments conducted over extended periods or through real time image acquisition in smaller frames for short periods of time. The potential for using multi-dimensional confocal microscopy as a powerful tool in cellular biology is continuing to grow as new laser systems are developed to limit cell damage and computer processing speeds and storage capacity improves.

Additional advantages of scanning confocal microscopy include the ability to adjust magnification electronically by varying the area scanned by the laser without having to change objectives. This feature is termed the **zoom factor**, and is usually employed to adjust the image spatial resolution by altering the scanning laser sampling period. Increasing the zoom factor reduces the specimen area scanned and simultaneously reduces the scanning rate. The result is an increased number of samples along a comparable length, which increases both the image spatial resolution and display magnification on the host computer monitor. Confocal zoom is typically employed to match digital image resolution with the optical resolution of the microscope when low numerical aperture and magnification objectives are being used to collect data.

Digitization of the sequential analog image data collected by the confocal microscope photomultiplier (or similar detector) facilitates computer image processing algorithms by transforming the continuous voltage stream into discrete digital increments that correspond to variations in light intensity. In addition to the benefits and speed that accrue from processing digital data, images can be readily prepared for print output or publication. In carefully controlled experiments, quantitative measurements of spatial fluorescence intensity (either statically or as a function of time) can also be obtained from the digital data.

Disadvantages of confocal microscopy are limited primarily to the limited number of excitation wavelengths available with common lasers (referred to as **laser lines**), which occur over very narrow bands and are expensive to produce in the ultraviolet region. In contrast, conventional widefield microscopes use mercury or xenon based arc-discharge lamps to provide a full range of excitation wavelengths in the ultraviolet, visible, and near-infrared spectral regions. Another downside is the harmful nature of high-intensity laser irradiation to living cells and tissues (an issue that has recently been addressed by multiphoton and Nipkow disk confocal imaging). Finally, the high cost of purchasing and operating multi-user confocal microscope systems, which can range up to an order of magnitude higher than comparable widefield microscopes, often limits their implementation in smaller laboratories. This problem can be easily overcome by cost-shared microscope systems that service one or more departments in a core facility. The recent introduction of personal confocal systems has competitively driven down the price of low-end confocal microscopes and increased the number of individual users.

Laser Systems for Confocal Microscopy

The lasers commonly employed in laser scanning confocal microscopy are high-intensity monochromatic light sources, which are useful as tools for a variety of techniques including optical trapping, lifetime imaging studies, photobleaching recovery, and total internal reflection fluorescence. In addition, lasers are also the most common light source for scanning confocal fluorescence microscopy, and have been utilized, although less frequently, in conventional widefield fluorescence investigations.

Lasers emit intense packets of monochromatic light that are coherent and highly collimated to form a tight beam with a very low rate of expansion. Compared to other light sources, the extremely pure wavelength ranges emitted by the laser have a bandwidth and phase relationship that is unparalleled by tungsten-halogen or arc-discharge lamps. As a result, laser light beams can travel over long distances and can be expanded to fill apertures or focused to a very small spot with a high level of brightness. Beyond the similarities common to all lasers, which include a gain medium (light source), excitation source (power supply), and resonator, these light sources differ radically in size, cost, output power, beam quality, power consumption, and operating life.

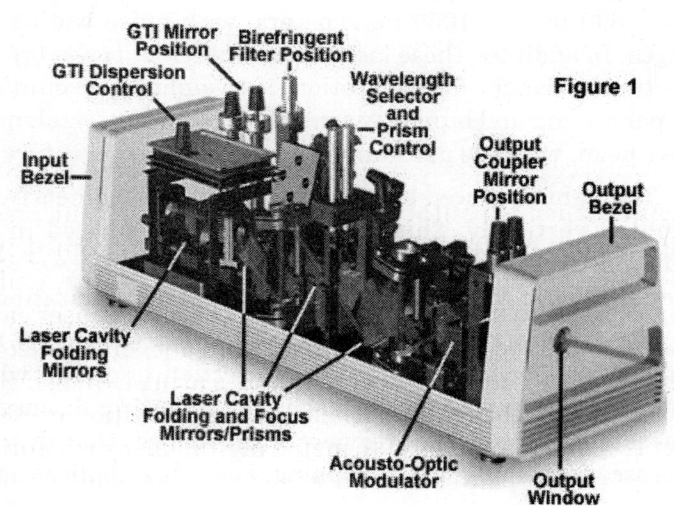

Fig. 7.42. Spectra Physics Tsumani® Tl: Sapphire Mode-Locked Laser.

The coherence of monochromatic light produced by most laser systems introduces problems in the application of these light sources for classical widefield microscopy. Light scattering and diffraction patterns are introduced by interference at every surface in the optical path. In addition, the field and aperture diaphragms, as well as dirt, also produce artifacts. These undesirable effects can be minimized or eliminated by a variety of techniques. The most common methods include temporally scrambling laser light by rapidly varying the optical path length between the light source and the microscope, or scanning the specimen point by point as is the case in confocal microscopy systems. In addition, interference and other artifacts can often be eliminated by the aperture scanning technique. If the path length or coherence state of the laser beam fluctuates at a faster interval than the detector integration time (in effect, the video frame rate), the speckle and scattering artifacts disappear from the image.

A successful technique employed by some investigators to improve differential interference contrast (**DIC**) images produced with an argon ion laser light source is to position a circular glass wedge, spinning at 2500 revolutions per minute, in the light path. Rapid variations in optical path length are introduced by differences in wedge thickness as the wedge rotates in front of the expanded laser beam. Currently, path length variation is usually accomplished by employing a fiber-optic light pipe to route the light between the light source and the microscope. Vibrating the fiber produces continual changes in the optical path length, causing the beam to become temporally incoherent at frequencies below that of the vibration level. The vibrations may be generated by a piezoelectric device, a loudspeaker, or the cooling fan utilized in the laser head.

Illustrated in Fig. 7.43 is a self mode-locked Ti:sapphire pulsed laser, which is currently one of the preferred laser excitation sources in a majority of multiphoton fluorescence microscopy investigations. Ti:sapphire mode-locked lasers provide a large wavelength tuning range, from about 690 to over 1050 nanometers, with pulse widths approximately 100 femtoseconds in length. In addition, these lasers have sufficient power (greater than 100 milliwatts throughout the tuning range) for saturation of two-photon excitation in most fluorophores. To ensure proper cooling and humidity control of the laser crystal, nitrogen gas is pumped into the sealed laser head, which is maintained at constant temperature by an external chiller.

The light produced by many laser systems is linearly polarized, with a polarization vector oriented vertically. This property can be exploited in applications requiring a polarized illumination source, such as differential interference contrast, polarized light measurements, or quantitative investigations of fluorescence polarization anisotropy.

The coherence and polarization characteristics of a laser beam are measured by the distribution of light in the beam cross-section, or profile, which changes with increasing distance from the exit mirror of the laser. The following discussion of laser beam characteristics is presented as a general overview of the subject that may prove useful in employing lasers in microscope imaging, laser trapping, and other applications.

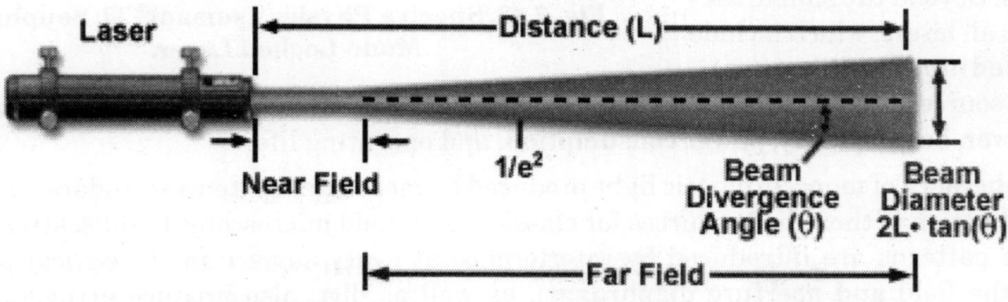

Fig. 7.43. Laser Beam Divergence in the Near and Far Field.

When a laser operates in the simplest **transverse electromagnetic mode**, referred to as the **TEM(00)** mode, the emitted beam has a planar wave front and a Gaussian intensity (irradiance) profile. The laser beam diameter is commonly defined as the value at which the intensity has fallen to **e(E-2)** (13.5 per cent) of its peak value. The Gaussian profile of the laser beam arises because of diffraction, which prevents the propagation of a perfectly collimated beam and induces transverse spreading of the light waves. Near the laser output aperture (termed the **near field**), the phase fronts of the beam can become disordered. As a consequence, the beam cross-sectional shape, size, and irradiance profile then change rapidly with distance from the laser. At greater distances (the **far field**), the phase fronts stabilize into the resultant Gaussian profile. In some literature references, the near field and far field are referred to by the alternative terms of **Fresnel zone** and **Fraunhofer zone**, respectively. The near field is also sometimes termed the **Rayleigh** range. The far field begins at a distance, **z**, defined by

$$z = A_0^2/l$$

where **A(0)** is the beam diameter at the exit aperture and **l** is the wavelength of light emitted by the laser. Applying this equation to an argon laser emitting a 0.6-millimeter-waist-diameter beam at a wavelength of 488 nanometers, the far field begins at approximately 74 centimeters from the exit aperture.

As discussed above, the beam can be considered essentially a parallel bundle of wavefronts that undergoes little spreading in the near field. Beyond the near field, the beam divergence angle (**q**), which is measured from the center of the beam to the edge (**e(E-2)**), grows larger and becomes the critical parameter in determining beam diameter (**D**) according to the equation:

Beam Diameter (D) = 2L • tan(q)

where **D** is the variable signifying laser beam diameter and **L** represents the length of the distance from the laser exit aperture to the measurement point on the beam. In practice, several laser beam characteristics, including the irradiance profile, are critical factors in many microscopy applications, and knowledge of the distance to the far field may be necessary in configuring the imaging system. Table 7.5 presents calculated values of this distance (using the equation given above) for a number of commonly utilized lasers and emission lines, and typical beam waist diameters.

Whether or not the beam exhibits Gaussian character is important in most laser applications because the beam often has to be focused, shaped, and otherwise modified by lenses and other optical components. A Gaussian beam has certain definable transformation characteristics, and these enable assumptions to be made about how the beam will propagate through an optical system.

The angular radius, designated by **q** (in radians), of a Gaussian beam in the far field is approximated by the expression:

$$\mathbf{q = l / pa_0}$$

where **a(0)** is the beam waist radius at the laser exit aperture. The beam waist diameter is a function of laser wavelength, cavity length, and other design parameters of the cavity. As the distance (**z**) from the laser increases, the beam waist radius is given by the equation:

$$\mathbf{a(z) = q_z}$$

Table 7.5. Distance to the Far Field.

Wavelength (nm)	*Beam Diameter (mm)*	*Far Field Distance (cm)*
	Argon-Ion	
488	0.6	74
514	1.0	195
Helium-Neon		
543	0.4	30
594	0.7	83
612	0.7	80
632	0.7	78
Nd:YAG		
355	3.0	2535
532	1.0	188
Ti:Sapphire		
790	2.0	506
395	2.0	1012
790	0.8	81

Typically, laser beams are characterized by beam **propagation** parameters such as the square of **M**, or **K** (which is equivalent to the reciprocal of the square of **M**), determined from a combination of near field and far field measurements as follows:

$$\mathbf{M^2 = pA_0q/4l}$$

Smaller values of $\mathbf{M^2}$, which is termed a **propagation constant** or **propagation factor**, are indicative of higher beam quality, particularly in reference to a smaller diameter and divergence. The factor describes the relationship of the real beam to that of an ideal Gaussian beam.

Coherent Gaussian beams have specific properties that cause them to differ from incoherent light beams in their propagation and transformation by lenses and mirrors. In the case of a diffraction-limited beam, the intensity profile of a Gaussian beam is itself Gaussian, provided that the beam is not truncated by the lens aperture. When the Gaussian beam diameter is one-half the aperture diameter of the lens, the intensity profile of the emergent beam remains Gaussian. When the Gaussian beam diameter is equal to the diameter of the lens aperture, the output beam intensity profile is a mixture of the Gaussian function and that of an Airy disk. Finally, a Gaussian beam diameter significantly larger than the diameter of the lens aperture produces the output profile of an Airy disk. In the latter instance, much of the laser power may be lost due to overfilling of the lens entrance aperture.

The overall topic of Gaussian beam optics is thoroughly covered in numerous textbooks, and details not discussed here may be obtained from more comprehensive sources. Two types of manipulation of the Gaussian beam are of particular interest to the microscopist utilizing lasers - **beam concentration** and **beam expansion**.

When a laser beam is focused to a very small spot by an aberration-free microscope objective (beam concentration), the radius of the spot at the focus (at distance **z**) is given by the expressions:

$$\mathbf{a(z) = lf/pa_0}$$

where *f* is the focal length of the lens. As an example, if a 100x objective having a numerical aperture of 1.3 (producing a focal length of approximately 1.6 millimeters) is employed to concentrate the 488-nanometer beam of an argon laser having a 0.3-millimeter radius, the focused spot radius (determined from the previous equation) is 0.8 micrometer. Increasing the beam waist fivefold through beam expansion (as discussed below) would result in a focused spot radius of approximately 0.16 micrometer.

It is important to note that extremely high power densities are achieved at the focal point of a concentrated laser beam. A 10-milliwatt beam focused to a diffraction-limited spot 0.22 micrometers in diameter results in a power density of approximately 30-million watts per square centimeter. Such high energy levels can rapidly degrade or destroy lens and filter coatings, as well as introduce considerable photochemical damage to biological specimens. However, for such a minute spot size, diffusion of thermal energy can be so effective in water that a high-energy, near-infrared beam may do little damage to a biological specimen unless absorption of the energy by the specimen is sufficiently high.

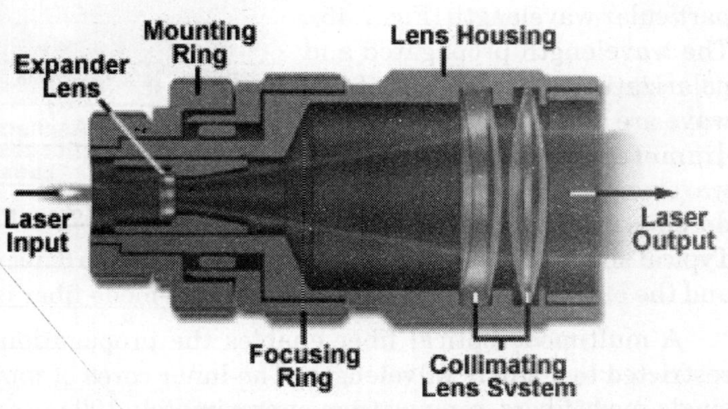

Fig. 7.44. Laser Beam Expander.

In many applications of lasers in optical microscopy, the laser beam is initially expanded by the use of a **Keplerian** or **Galilean** beam expander, either of which is actually a reversed telescope (typical laser beam expander anatomical features are illustrated in Fig. 7.44). The divergence of a coherent Gaussian beam can be reduced, and the beam optimally collimated over a longer distance, if the laser beam is first expanded. Referring to the previous equations, the angular radius of the beam, designated **q**, is inversely proportional to the beam waist radius, **a(0)**, at the laser exit aperture. Therefore, expanding the beam waist radius diminishes the divergence proportionally.

It is practical for many applications in microscopy to pipe the laser output directly into the microscope optical path through a flexible optical fiber (as illustrated in Figure 94). This technique is preferable to the alternative method of rigidly aligning the laser and microscope, which requires employing a massive, vibration-free optical table and numerous fixed mirrors and other components.

When a laser beam is focused by a lens onto an optical fiber, the coupling efficiency and characteristics of the beam that emerges from the fiber depend heavily on the fiber geometry. Most of the optical fibers utilized for laser light delivery are constructed with a fused-silica core. These fibers consist of an inner core fabricated with high-refractive-index silica and surrounded by a sleeve, termed the **cladding**, composed of lower-refractive-index material. Light is prevented from escaping the fiber along its length by total internal reflection at the interface of the core and the cladding. The cladding may be composed of silica, glass, a hard fluoropolymer, or a soft silicone.

Optical fibers are classified as **single-mode** or **multimode** according to the diameter of their inner cores. A single-mode fiber allows propagation of only the lowest-order mode at one particular wavelength (Fig. 7.45). The wavelength propagated and polarization preservation of the wave are determined by the fiber diameter. Although other wavelengths may propagate, they do so with reduced efficiency. Typical single-mode fiber diameters range from 3 to 6 micrometers for visible-light wavelengths, and the output irradiance profile of a single-mode fiber is Gaussian.

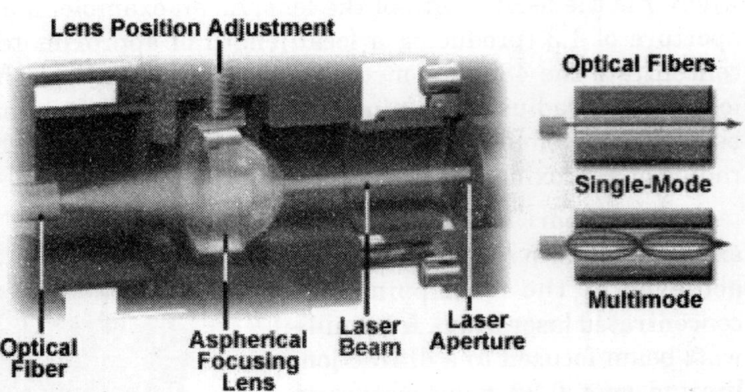

Fig. 7.45. Fiber Optic Laser Coupler.

A multimode optical fiber enables the propagation of more than one mode, and is not restricted to a single wavelength. The inner cores of multimode optical fibers are larger than single-mode fibers, ranging from approximately 100 micrometers to 1.2 millimeters in diameter. The output irradiance profile from a multimode fiber has a flat shape, referred to as a **top-hat profile**, with a numerical aperture that is determined by the core and cladding refractive indices.

The acceptance cone angle, **q**, of the fiber core is related to the numerical aperture, **NA**, of the fiber as follows:

$$\mathbf{NA = \sin q/(n^2{}_{core} - n^2{}_{cladding})^{1/2}}$$

where **n** represents refractive index. Efficient coupling of laser light to the fiber core occurs when the fiber core numerical aperture and that of the beam concentration lens are matched. The efficiency of light transmission through an optical fiber is typically as high as approximately 90 per cent, but may be sharply reduced (to only 60 or 70 per cent) by bends having very small radii (less than 3 centimeters).

In utilizing any laser, it is crucial not only to prevent any direct or specularly reflected laser light from entering an observer's eyes, but also to avoid reflection of the beam from a component of the optical system back into the laser system. The former is an obvious personal safety precaution, while the latter caution is important to prevent an additional reflector from returning a coherent beam back into the laser, causing possible damage to the system.

Stability of the laser light source is an important aspect in many applications, particularly in quantitative microscopy, where illumination intensity fluctuations can adversely affect experimental results. A number of factors related to stimulated emission and cavity length fluctuations can induce frequency noise in the output beam, but other perturbations causing amplitude fluctuations can create both high-frequency **intensity noise** and slow variations (drift) in optical output power. Some sources of these intensity fluctuations are related to the function of the laser head itself or the power supply. The most common sources of noise in the output beam for various laser categories are listed below:

- **Gas lasers** - Mirror misalignments from resonator vibrations, noise from optical pump sources, plasma oscillations and instabilities of the ion discharge process, fluctuations in power supply current, microphonics from cooling water turbulence, and fan-induced noise in forced-air cooling systems are all potential noise sources.
- **Solid-state lasers** - Noise sources include microphonics, pump source fluctuations for both lamp and diode pumps, cavity alignment errors, and the random frequency-related noise (termed ***1/f*** noise) that is related to thermal fluctuations in the laser medium.
- **Dye lasers** - Both noise (high-frequency) and drift result from density inhomogeneity and air bubbles in the dye solution, and by dye pump and laser pump source instabilities.
- **Semiconductor (diode) lasers** - Noise can result from fluctuations in the drive (bias) current or temperature, and **1/f** noise is caused by trapping of carriers in the junction and by other types of carrier (electron-hole) recombination effects.

All lasers are susceptible to noise introduced by their power supplies. **Switching** power supplies, which have become common because of their efficiency and small size, are particularly likely to introduce ripple to the laser system at frequencies ranging into the tens of kilohertz. Such interference, when it affects the light beam in optical microscopy systems, can be especially troublesome to diagnose and remove. The primary difficulty is due to the similarity with noise introduced into the system by other sources, such as electromagnetic fields in the laboratory environment. In order to achieve adequate output stability, semiconductor lasers must be operated with diode current supplies having the highest electrical stability and lowest noise available, and with precise temperature control. Other external noise sources must be controlled, including dust in the laboratory, and vibrations originating from local traffic and building equipment.

The beam intensity of continuous wave (**cw**) lasers can be stabilized by either electronic control of the tube current or through utilization of external components that modulate the light intensity. Two different methods are often employed to control the tube drive current. In the **constant current** mode, tube current is directly controlled by an electronic feedback loop to minimize fluctuations. Because the laser output is also temperature dependent, this type of control circuit is most effective if adequate temperature control is provided. **Constant output power** stabilizing systems operate by controlling the drive current in response to a signal derived from a circuit that samples the output beam using a beamsplitter and photodiode monitor. This physical arrangement is applicable for gas lasers and several other geometries, but smaller diode lasers are commonly assembled in a package that already includes an integral photodiode. The monitor photodiode samples emission from the rear facet of the laser wafer and produces a signal that enables feedback control of the output power.

External components utilized to provide stabilization of laser intensity generally employ a fast feedback system to control an electro-optic modulator that minimizes fluctuations in beam power. The external **Pockels cell** modulator (see Fig. 7.46) is available from a number of manufacturers, and can be used, in principle, to stabilize the output power of any continuous wave laser. Large intensity fluctuations (up to approximately 50 per cent) can be corrected by this technique, but with a proportional reduction in total output power. A wide range of correction capability is important with many systems. The helium-cadmium laser, for example, can exhibit variation in output power of around 20 per cent, due in part to strong plasma oscillation between certain beam frequencies. Systems exist that are reported to be suitable for regulating **cw** and **mode-locked** lasers to within a few hundredths of a per cent of their output power, and over a frequency range from direct current to several hundred megahertz, with noise attenuation of 500:1, or greater.

The basic components of a Pockels cell modulator are presented in Fig. 7.46. External devices for regulating laser output intensity, similar to the design illustrated in Fig. 7.46, are sometimes categorized or marketed under the term **noise eaters**. The fundamental concept behind electro-optic modulators utilizing the Pockels effect is based on a mechanism for varying the polarization properties of the cell at an extremely rapid rate to provide a variable beam attenuator for control of laser intensity. The polarization state of the laser output determines the total attenuation of the modulator, but up to 80 per cent transmission is possible. Following emission from the laser head, a portion of the beam is diverted by a beamsplitter to a photodiode, which compares the intensity to a preset (selectable) reference intensity, and amplifies the difference signal so that it can drive the electro-optic Pockels cell modulator. The amplified signal produces a refractive index change in the cell that rotates the plane of polarization, thereby varying the beam attenuation in proportion to the difference voltage that is applied. Among substances that exhibit a change in polarization properties with changes in electric field (Pockels effect) are potassium dihydrogen phosphate and lithium niobate, and crystals of these materials are commonly utilized in the beam modulators.

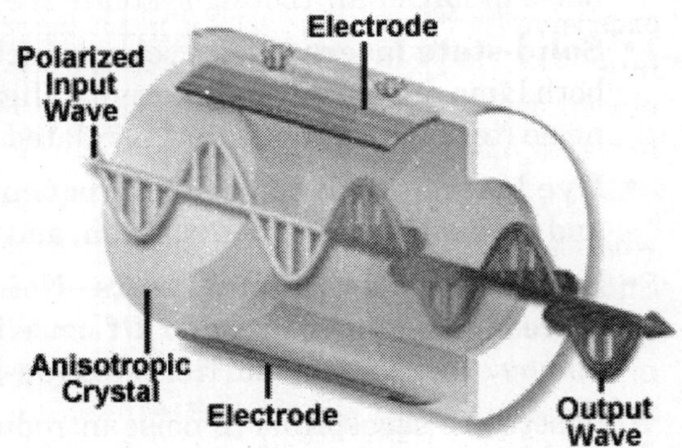

Fig. 7.46. Anatomy of the Pockets Cell.

In situations where randomly polarized light is stabilized by a Pockels cell system, the modulator must be positioned between crossed polarizers, and further consideration is necessary to minimize the effects of these additional components on the beam stability. Because dust, vibrations, and other interferences can alter beam stability at any point in the optical path, it is important that external stabilizers are placed as close as possible to the specimen position in optical microscopy systems. This effort will ensure that the most stable beam is delivered to the specimen.

Both the **argon-ion** laser and **krypton-ion** laser produce multiple emission lines from rare-gas transitions that differ substantially in power levels, and only a few of the lines are suited for microscopy applications. The air-cooled argon-ion laser is widely employed as a light source for confocal microscopy because of its brightness level, small size, excellent beam geometry, and the suitability of its spectral lines for fluorescein and (less efficiently) rhodamine excitation. Most argon-ion lasers utilized in widefield or confocal fluorescence microscopy emit only two usable lines, 488 and 514.5 nanometers, which represent approximately 75 per cent of the total laser power. Higher-power (greater than 5 watts) argon-ion lasers equipped with special mirrors can emit ultraviolet lines at 334, 351, and 364 nanometers, and additional lines at visible wavelengths extending from 458 to 529 nanometers.

Krypton-ion lasers have found fewer applications in microscopy than argon lasers due to their somewhat longer wavelength output. In addition, krypton produces only 10 to 30 per cent as much power as argon when used in the same tube, and often requires water-cooling to generate the equivalent power output of an air-cooled argon system. A major drawback of air-cooled ion lasers is their lack of efficiency, which results in large power requirements and excessive heat generation that must be removed from the system by forced air with exhaust fans. The lifespan of ion lasers is reduced due to gas consumption, with trapped gas being buried within the walls of the discharge tube as a result of the high current densities necessary for laser operation.

Air-cooled lasers using argon-krypton mixtures have become popular in confocal microscopy when several illumination wavelengths are required for dual or multiple-fluorophore studies. Such **mixed-gas** lasers are only capable of producing stable output on major lines that are well separated in the wavelength spectrum. Of the three laser lines typically utilized for confocal microscopy, the 488-nanometer and 568-nanometer lines have approximately equal power (10 to 15 milliwatts), while the 647-nanometer line has about 50 per cent more (15 to 25 milliwatts). All ion lasers exhibit excellent beam quality and can be purchased from a variety of manufacturers in single-line, multi-line, and tunable configurations.

The emission at 633 nanometers (termed the **He-Ne line**) of the common **helium-neon** laser has been supplemented by development of variants having emissions in the green (543 nanometers), yellow (594 nanometers), orange (612 nanometers), and near infrared (1523 nanometers) spectral ranges. Although most of these lasers are single-line emitters and relatively low in power (less than 10 milliwatts), the helium-cadmium laser is an exception, emitting at 325 or 442 nanometers with greater than 50 milliwatts of power.

Helium-cadmium lasers (illustrated in Fig. 7.47) are considered to be a member of the helium-neon family, and represent an economical source for continuous-wave output in the ultraviolet (325 nanometers at 75 milliwatts and 353 at 20 milliwatts) and violet (442 nanometers at 200 milliwatts) spectral regions. These lasers rely on cadmium vapor as a lasing medium, which is distributed uniformly throughout the bore from the heated cadmium reservoir (at approximately 250 degrees Celsius) by gas-phase electrophoresis. The helium pump maintains a constant helium pressure, which is about a thousand times higher than that of the cadmium vapor. Helium-cadmium lasers display more beam noise than their helium-neon counterparts, primarily because of localized fluctuations in metallic cadmium vapor concentrations in the bore. They also have a shorter lifespan, typically around 5,000 operating hours.

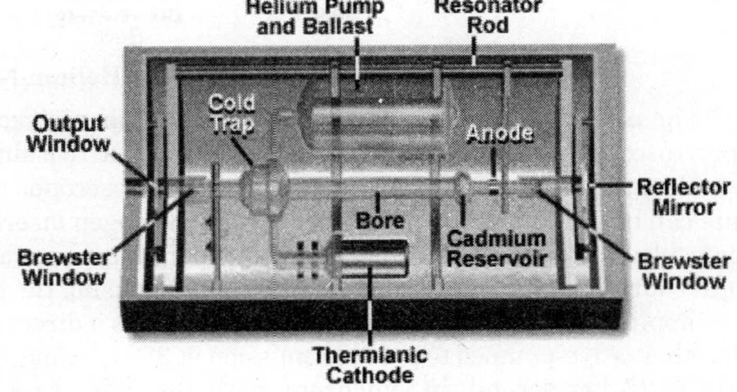

Fig. 7.47. Helium-Cadmium Laser Internal Components.

One group of investigators has described the use of a ratio-imaging confocal microscope coupled to both an argon laser operating at 488 nanometers and a helium-cadmium laser emitting at 442 nanometers. This system allowed the group to measure the intensity ratio from the pH-sensitive dye **BCECF** [having an **IUPAC** name of 2',7'-bis-(2-carboxyethyl)-5-(and-6-)-carboxyfluorescein] in isolated renal tubules. Similar measurements have been made utilizing the 458-nanometer line of the argon laser in both widefield and confocal microscopy systems. Another researcher reported using the 325-nanometer output of a helium-cadmium laser to excite two dyes simultaneously and perform emission ratio imaging.

Helium-neon lasers are the most widely utilized laser systems for a broad range of biomedical and industrial applications, and display the most superior Gaussian beam quality of any laser. These lasers are readily available at relatively low cost, have compact size dimensions, and exhibit a long operating life (often reaching 40,000 to 50,000 hours). The low power requirements, superior beam quality (virtually a pure Gaussian profile), and simple cooling requirements (convection) make helium-neon lasers the choice system for many confocal microscopes.

Presented in Fig. 7.48 is a cut-away diagram of a typical helium-neon laser system, which is constructed of glass with a large oxidized-aluminum **cold cathode** as the electron emitter. Operating in the **abnormal glow** current density gas discharge region, helium-neon lasers are generally high-voltage and low current systems, with discharge currents being limited to a few milliamperes and potentials ranging from several hundred to a thousand volts. Progressive deterioration of the oxide coating on the cathode, which ultimately leads to sputtering of aluminum, is the limiting factor in helium-neon laser operating life. Large-diameter discharge tubes typically have longer life spans than smaller tubes (40,000 hours versus about 10,000 hours, respectively).

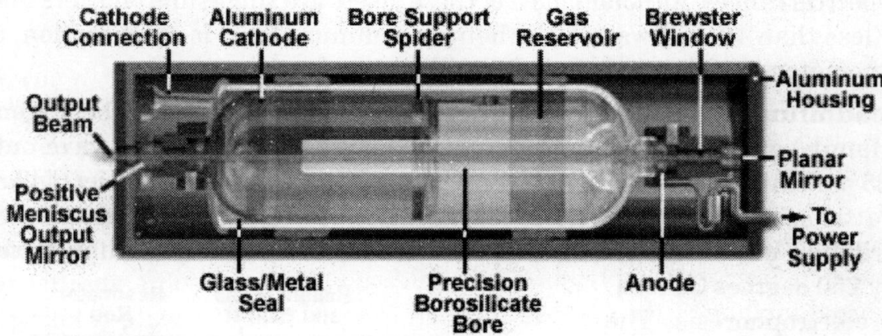

Fig. 7.48. Anatomy of the Helium-Neon Laser.

The **nitrogen** laser has been in use for a number of years as a pulsed light source for both spectroscopy and microscopy. The output is confined to a single line, having a 337.1-nanometer wavelength, with pulse durations ranging from picoseconds to nanoseconds. The pulse repetition rate can be as high as 200 pulses per second. Nitrogen lasers can also be employed to pump dye molecules for longer-wavelength emission. In addition, these lasers have been utilized as a light source for high-speed calcium ratio imaging using the ultraviolet-excitable dye **Fura-2**. In this application, two lasers are used; one serves as a direct source of 337-nanometer light, and the other is dye-pumped to produce emission at 380 nanometers. Pulsing each laser at a repetition rate of 15 per second, in synchrony with the video rate, produces a ratio image every 66 milliseconds.

Diode lasers, semi-conductor devices that have been under development for decades, are now available with sufficient output power to be of interest to microscopists. The most common of these devices operates in the near infrared, but diode lasers operating in the red and blue regions (and recently, other wavelengths), with substantial output power, have been developed. In addition, those diode lasers now available exhibit improved beam shape and stability, allowing them to replace helium-neon lasers in many applications. Diode lasers typically have a lifespan ranging between 10,000 and 50,000 hours, but are extremely sensitive to electrostatic shock, so they must be handled carefully.

Of great interest to optical microscopists is the development of tunable diode lasers, which can now compete in terms of power and versatility with tunable dye lasers and Ti:sapphire lasers (discussed below and illustrated in Figure 7.42). Tunable dye lasers have a wavelength range of 600 to 1800 nanometers and can deliver 5 to 25 milliwatts of power. They have the advantages of relatively low cost, compact size, long lifetime, and low heat production, eliminating the requirement for external cooling systems.

Diode-pumped solid-state lasers (**DPSS**) utilize a diode laser instead of noble gases, arc lamps, or flashlamps to pump the solid-state lasing material. The power output, beam quality, and stability exhibited by diode-pumped lasers approaches that of a gas (helium-neon) laser, but the efficiency and size are more comparable with diode lasers. Typical operating and maintenance costs of diode-pumped lasers is less than that of gas lasers, and most systems are cooled either by convection or forced air.

Diode-pumped neodymium-yttrium aluminum garnet (**Nd:YAG**) lasers generate 1064-nanometer light in the milliwatt power range. Frequency doubling leads to a compact device with a continuous-wave output at 532 nanometers, and frequency tripling can also be employed to generate a pulsed output at 355 nanometers. The tightly folded resonator (**TFR**) was developed for pumping a crystal of neodymium-yttrium lithium fluoride (**Nd:YLF**) with high power and efficiency using an array of diode lasers to generate several watts of power at 1047 nanometers. Frequency doubling, tripling, and quadrupling in this type of laser results in power outputs up to hundreds of milliwatts of coherent light at 523, 349, and 262 nanometers (second, third, and fourth harmonics). Other advantages of diode lasers as pump sources include an extended lifetime (typically more than 5000 hours, compared to a few hundred hours for lamps), a collimated and easily focused output that matches the small lasing volume of the solid-state laser, and greatly reduced thermal loading of the laser rod, which usually requires water cooling when halogen arc lamps are used as pumps.

Development of diode-pumped solid-state lasers has been driven by industrial and commercial applications requiring high power (generally several watts) in the green (532 or 523 nanometers) and ultraviolet (355 or 349 nanometers, and the fourth harmonic at 266 nanometers) wavelength ranges. Output in the ultraviolet spectral region is pulsed, with energies ranging from 100 microjoules to 10 millijoules, pulse durations in the nanosecond range, and repetition rates as high as 10 kHz. These lasers are very useful in microscopy for triggering the release of caged compounds. The pulse repetition rates, however, are still too slow for use as an illumination source for most confocal microscopy applications.

Further development has led to the combination of diode-pumped solid-state lasers with **optical parametric oscillators** (**OPOs**; see Figure 7.49) to produce a tunable, pulsed output that is continuously variable from 205 nanometers to 2000 nanometers. Although the initially-available systems have been expensive and complex in operation, scaled-down versions more suitable for use in microscopy are being introduced.

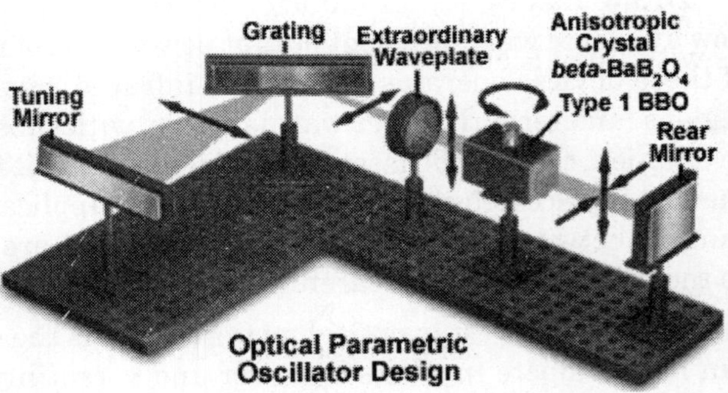

Fig. 7.49. Optical Parametric Oscillator Design.

Titanium-doped sapphire lasers (commonly known as **Ti:sapphire** lasers, see Figure 7.42) provide the advantages of tunability for pulsed and continuous light delivery, as well as solid-state dependability. These lasers can deliver very short light pulses (approximately 80 to 100 femtoseconds) at high repetition rates (100MHz). The range of tunable wavelengths extends from the far red to the near-infrared spectral regions (700 to 1000 nanometers). Most of these lasers are operated with optical pumping by high-power argon lasers, as well as requiring water cooling. As a result of the expense and complexity involved with operating and maintaining Ti:sapphire lasers, their use has been limited primarily to multiphoton microscopy in relatively few laboratories.

Recently, a diode-pumped **Cr:LiSAF** (chromium-doped lithium strontium aluminum fluoride) laser has been developed that delivers high-frequency 90 femtosecond pulses at a wavelength of 860 nanometers, and an average output power of 88 milliwatts. The small size and low power requirements of the Cr:LiSAF laser make it an attractive light source for multiphoton fluorescence microscopy in place of the Ti:sapphire laser.

Confocal Microscope Objectives

For any conventional optical microscope configuration, the objective is the most critical component of the system in determining the information content of the image. The contrast and resolution of fine specimen detail, the depth within the specimen from which information can be obtained, and the lateral extent of the image field are all determined by the design of the objective and its performance under the specific conditions employed for the observation. Additional demands are imposed on the objective in scanning confocal techniques, in which this crucial imaging component also serves as the illumination condenser and is often required to perform with high precision at a wide range of wavelengths and at very low light levels without introducing unacceptable image-degrading noise.

Regardless of the function of any other system component, no information can be added to the image that was not initially captured by the objective. Certain intermediate components in the imaging path may perform corrective functions, but their primary performance requirement is that they degrade the fundamental image information received from the objective as little as possible. Traditionally, the primary variables considered in objective selection for a particular application have been the magnification, whether a dry or immersion design is required, and the numerical aperture of the lens system. The development of new imaging techniques driven by technological improvements in lasers, fluorescent dyes, and new specimen labeling capabilities, as well as continuous optical improvements by the microscope manufacturers, have dramatically enhanced progress in certain research areas. This is particularly true in the fields of cell and molecular biology and neuroscience, and the confocal microscope combined with fluorescence techniques has come to be widely relied upon as a research tool.

High performance objective designed specifically for use with laser illumination in both the visible and infrared regions of the spectrum, presented in Fig. 7.50. The objective is a 60x flat-field (plan) apochromat water immersion model, which is corrected for wavelengths ranging from 450 to 1100 nanometers. Among the useful applications for this objective is simultaneous fluorescence and differential interference contrast (**DIC**) observations without suffering chromatic aberration between the visible fluorescence and infrared DIC images. The objective is also fitted with a correction collar that enables adjustment for fluctuations in cover glass thickness and reduction in spherical aberration when imaging deep within tissues or through aqueous solutions. Similar objectives with spherical and chromatic aberration correction in the 400-700 nanometer region are also available, as are models designed for ultraviolet imaging down to 350 nanometers. Several versions of the latter objective are corrected to bring both ultraviolet excitation and blue emission to the same focal point as visible light, thus enabling true confocal imaging throughout the viewfield.

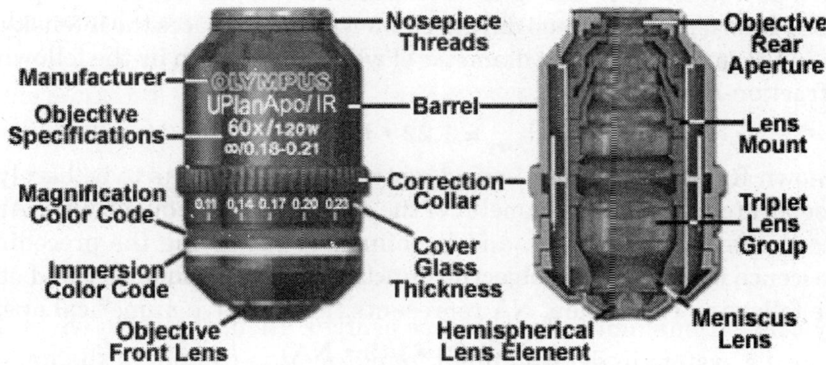

Fig. 7.50. High Performance Confocal Microscope Objective.

The general demands placed upon the objective in the confocal instrument are similar to those in other critical microscopy applications, although the ever-expanding requirements imposed by new techniques have made it more common than ever to approach the performance limits of these lens systems. The specific limitations of certain applications have made it apparent that other objective characteristics may be as important, or more important, than those traditionally considered paramount. In order to meet essential requirements of currently promising techniques, manufacturers have introduced new optics specifically aimed at optimizing performance for these methods. The development of high numerical aperture, highly corrected water immersion objectives is an example of response to the dramatic increase in living cell and tissue research, which is by necessity conducted in aqueous media. It is possible that specific requirements in confocal microscopy will lead to the development of optics that less stringently correct one or more of the resolution-compromising aberrations in favor of achieving other design goals that are more critical in confocal performance. A commonly-discussed example is the need for higher photon collection efficiency in living cell studies, which may provide an incentive for development of objectives that sacrifice correction of certain aberrations in order to maximize transmittance, especially in crucial fluorophore emission wavelength bands.

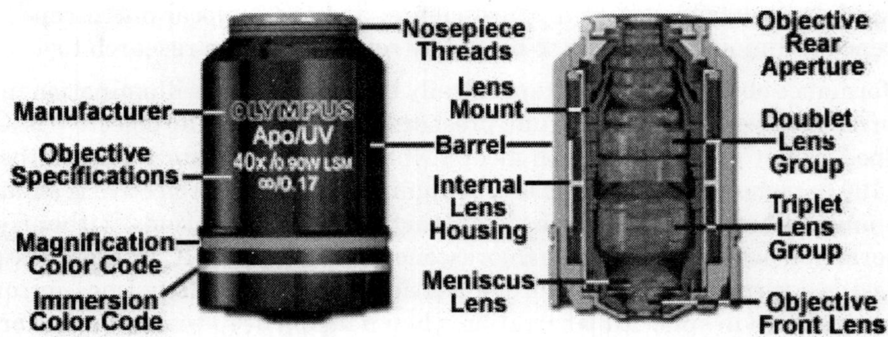

Fig. 7.51. Ultraviolet Confocal Microscope Objective.

Resolution in optical systems is commonly accepted to be the minimum separation distance between two specimen features that enables them to be distinguished as separate in the final image. Defined on the basis of a specified contrast required for visual recognition, resolution can be quantified as a function of the light wavelength (**l**) and the numerical aperture (**NA**) of the optical system. Interference between diffracted wavefronts produces the intensity distribution of the Airy disk in the image plane, the diameter of which (**d**) is given by the following equation, in the ideal diffraction-limited system:

$$\mathbf{d_{Airy} = 1.22 \bullet l/NA}$$

The well-known Rayleigh criterion considers two identical points to be barely discernible when they are separated by half the diameter of the Airy disk. Consequently, the Airy radius (**r**) is equivalent to the lateral resolution, and is defined by modifying the preceding equation. When epifluorescence is utilized, the objective functions as both condenser and objective, and therefore in the following expression, **NA** represents the objective numerical aperture:

$$\mathbf{r_{Airy} = 1.22 \bullet l/(2 \bullet NA)}$$

Diffraction limitations that result in lateral redistribution of intensity from point objects into Airy patterns also produce defocusing along the optical axis, which changes the size and shape of the image point. The three-dimensional intensity point spread function describes the intensity distribution and represents the performance of the optical system in transferring specimen information to the image plane. A range of optical aberrations are characteristic of refractive lens systems, and any that are not corrected by the objective design or compensated by another optical component will alter the intensity distribution that represents each specimen point in the image, degrading the image with respect to the ideal diffraction-limited performance. The relative importance of the potential aberrations in confocal applications must be taken into consideration in selecting objectives for the technique. The principal factors in objective design and performance, as they pertain specifically to the confocal microscope, are discussed in the following sections.

Currently, the most widely used confocal microscope configuration for biological applications employs a computer controlled scanning system to deflect an illuminating laser beam, focused through the objective, across a stationary specimen. This is referred to as **off-axis** confocal scanning because the scanned beam is deflected to each side of the optical axis and utilizes peripheral regions of the objective lens elements. Some early implementations of scanning techniques in the optical microscope utilized **on-axis** scanning in which the laser beam remained fixed on the optical axis and the specimen stage or the objective was scanned.

Although in principle the two scanning methods can produce similar results, the requirements with regard to choosing an objective for optimum performance are significantly different for on-axis and off-axis illumination scanning, and the considerations for collection of emitted fluorescence through the objective differ as well. A third scanner type utilizes a spinning **Nipkow** disk to scan numerous illumination points over the specimen without moving the light source or microscope stage. Although this methodology is widely used in the semiconductor industry and can allow real time reflected-light confocal viewing, it is less commonly applied in biological instruments. Nipkow systems are likely to become increasingly important in live cell investigations, however, because of their potential to minimize cell damage compared to configurations employing high-intensity scanned laser sources.

Lens aberrations that affect objective performance may be divided into two categories, including those that are non-chromatic (invariable with wavelength), and the chromatic aberrations, which are wavelength dependent. The chromatic aberrations are characterized as either **lateral chromatic aberration** or **longitudinal chromatic aberration**, and the wavelength independent group includes **spherical** aberration, **coma**, **astigmatism**, **field curvature**, and **distortion**. Chromatic aberrations and spherical aberration affect the entire image field, while the remaining aberrations are more prevalent in off-axis zones. The chromatic and spherical aberrations (see Figure 7.52) are the most likely to negatively affect confocal performance. In general, chromatic aberrations cannot be modified procedurally and must be dealt with by design of the optical components. Other artifacts, especially spherical aberration, are commonly exacerbated by improper use of the objective or introduction of mismatched optical components, but can be minimized or compensated by following the appropriate techniques or with adjustments to the optical system.

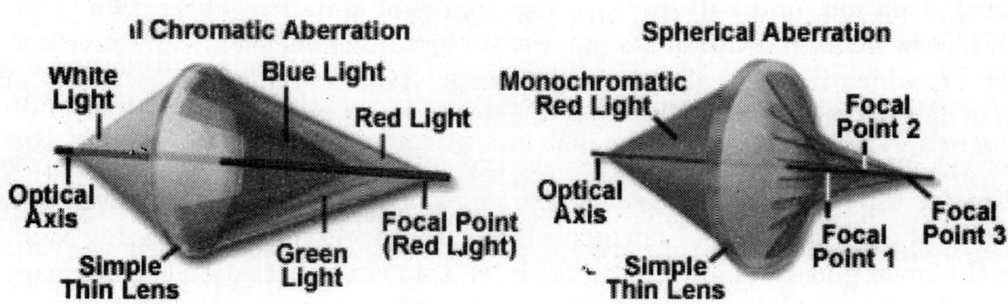

Fig. 7.52. Common Lens Aberrations in Confocal Microscopy.

Spherical Aberration

Spherical aberration is the non-chromatic aberration that is most significant to confocal performance, and is a manifestation of the property of spherical lens components that causes paraxial and peripheral light rays to be focused in successive planes. A varying degree of refraction for ray paths through different lens zones results in blurring of the focused image of a point source, and produces an asymmetrical intensity change above and below the focal plane. In all modern objectives, spherical aberration is corrected to visually imperceptible levels if the specified operating variables for the objective design are exactly satisfied. Unfortunately, several possibilities exist in practice to allow deviation from the optical design criteria of the objective, and to induce spherical aberration. Because spherical aberration can only be adequately corrected for a precisely specified distance relationship between the objective and the specimen and image planes, the artifact can inadvertently be introduced if the tube length specified for the objective is not maintained. This can occur if the objective is used on a microscope having other than the required tube length in a finitely corrected system, or by the introduction of optical elements, such as filters, into the converging beam path in such a system.

Optimum correction of spherical aberration requires careful attention to the imaging media external to the objective, and this is another potential source of degraded performance. The objective characteristics are fixed by design and cannot be expected to accommodate a range of operating conditions (with the exception of adjustments provided by correction collars). Among the factors that can add significant spherical aberration are poor quality immersion oil between the objective and specimen, non-standard cover glass thickness, the specimen mounting medium, and the specimen itself. Any material that enters between the specimen and front lens surface of the objective becomes a critical component of the imaging system. Adherence to the design requirements of the objective becomes even more critical at higher numerical aperture. Variations in cover glass thickness or in its refractive index can increase spherical aberration, especially with "dry" (non-immersion) objectives. High numerical aperture dry objectives are generally designed for optimum performance with a cover glass of 0.17-millimeter thickness, having the specimen mounted directly beneath and in contact with the glass.

To allow for correction of spherical aberration when non-standard cover glass thickness is used, many objectives are equipped with an adjustable correction collar that can be set to a range of thickness settings. The correction collar operates by translating an internal lens group to alter the objective focal length. Even if a cover glass of the correct thickness is used, the presence of a layer of mounting medium between the glass and specimen deviates from the ideal optical situation, and will increase the degree of spherical aberration. The correction collar may also be utilized to minimize spherical aberration induced by refractive index changes of this nature, which in confocal microscopy causes reduction of intensity at the pinhole and reduction of depth discrimination because of axial focus shifts.

Objectives intended for oil immersion are typically optimized for use with a cover glass of 0.17-millimeter thickness having a refractive index of 1.518 at specified wavelengths, and an immersion oil of precisely defined refractive index. By specifying the operating conditions with respect to the cover glass and immersion medium, spherical aberration can be corrected by the objective design for several values of wavelength (depending upon the type of objective). The importance of matching the refractive indices between each material throughout the light path, from the specimen and mounting medium to the objective front lens element, has been historically one of the most problematic imaging criteria, especially in attempts to achieve high resolution with biological specimens. The refractive index of sub-cellular components is considerably lower than that of conventional immersion media, and in many cases these refractive indices are uncertain and vary throughout the specimen. Even in fixed materials, the mounting medium refractive index is typically not identical to that of available immersion oils.

For investigations of dynamic processes in living cells, which are cultured and maintained in physiological saline solution, the mismatch of oil and water refractive indices results in a critical limitation on the performance of oil immersion objectives. The primary problem is that the full potential of the highest numerical aperture oil immersion objectives is not achieved due to the mismatch between the refractive index of water (1.33) and that of immersion oil (approximately 1.5). A critical factor in the application of confocal microscopy is improved fluorescence imaging and three-dimensional representation of thick specimens, and the spherical aberration introduced when oil immersion objectives are used with aqueous specimens limits the depth into the medium at which acceptable images can be obtained. In general, high numerical aperture oil immersion objectives are designed to be used at image planes no more than 15 to 20 micrometers beneath the cover glass. When applied to aqueous specimens, however, the spherical aberration induced at the water-cover glass interface can reach substantial levels at depths as low as 10 micrometers. As an example, the point spread functions illustrated in Fig. 7.53 demonstrate the increasing degree of spherical aberration occurring with a plan apochromat oil immersion objective at penetration depths ranging from zero to 8 micrometers.

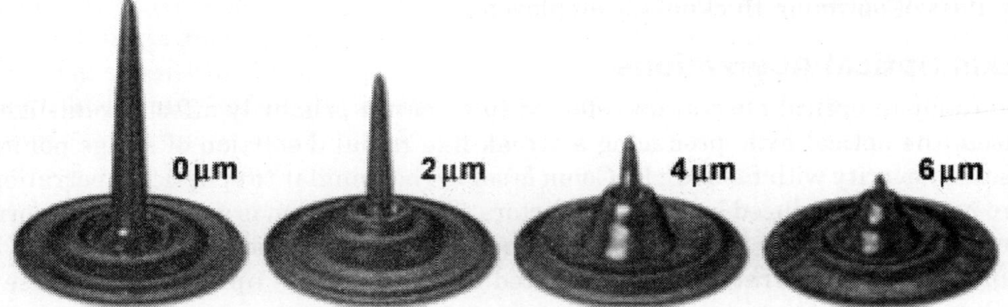

Fig. 7.53. Point Spread Function Intensity Distribution in Water.

The interest in utilizing confocal fluorescence techniques for collecting three-dimensional data from aqueous biological specimens has been a primary incentive for microscope manufacturers to introduce a number of high numerical aperture, highly corrected water immersion objectives. Spherical aberration is a major optical limitation in confocal studies of living cells when oil immersion objectives are employed, and the effects increase in proportion to the depth of observation into aqueous media and variable sub-cellular components (see Fig. 7.53). The negative effects on imaging include loss of contrast and signal intensity due to a reduction in the fraction of emitted fluorescence intensity reaching the detector pinhole, loss of resolution of minute specimen features, and reduction of **z**-axis position accuracy, which affects the integrity of three-dimensional image reconstructions. By utilizing water immersion objectives when imaging specimens in aqueous media at significant depth beneath the cover glass, the spherical aberration degradation can be avoided or minimized, allowing the full benefits of the confocal technique to be realized. Water immersion objectives having long working distances are capable of collecting accurate three-dimensional data at depths of more than 200 micrometers in aqueous media.

An oil immersion objective, regardless of the degree of optical correction, is not the optimum choice for imaging a water-immersed specimen. Optimum image quality is possible with this combination only for specimen regions that are in direct contact with the cover glass. At greater depth, the effects of spherical aberration degrade contrast and resolution and reduce image brightness to the extent that confocal signal-to-noise ratio is drastically reduced. The best solution for minimizing induced spherical aberration when imaging at large distances into aqueous media is use of a highly corrected water immersion objective. Although objectives designed for direct immersion of the front lens element into the specimen medium are available, the change of focus that is required in collecting three-dimensional data may produce undesirable movement of the specimen between images in a sequence. To avoid this specimen motion when the objective-specimen distance is varied, a cover glass must be used in most cases.

A number of the recently introduced high numerical aperture water immersion objectives incorporate adjustment collars that provide variable correction for spherical aberration, such as that induced by fluctuations in cover glass thickness. This type of correction collar is also useful in compensating for refractive index differences in physiological media and in cell and tissue constituents, and for variation of refractive index with temperature or solute concentration changes. Due to the variety of factors that can contribute to induced spherical aberration, adjustable correction is a desirable feature of an objective for confocal applications even when a cover glass of optimum thickness is employed.

Off-Axis Optical Aberrations

The common optical aberration referred to as **coma** primarily affects point-like sources away from the optical axis, producing a streak-like radial distortion of image points, which increases in severity with field angle. Coma is somewhat similar to spherical aberration in that both are potentially induced by the same factors. This aberration is generally well corrected in modern optical systems by the utilization of appropriate lens elements, and objectives in which coma and spherical aberration are eliminated are classified as **aplanatic**. Because coma is primarily an off-axis aberration, it has significance to confocal laser scanning systems, which commonly utilize off-axis beam paths, but is not a factor with specimen-scanning (on-axis) confocal microscopes.

Several additional types of geometric aberration are potentially significant to confocal objective performance, and because all are more evident in regions of the image field away from the center, they are more likely to affect performance in laser scanning confocal systems. The optical demands of confocal imaging are such that only highly corrected optics are likely to perform adequately, and significant geometric aberrations will typically have been minimized in objectives from this category. It is beneficial, however, to be aware of these aberrations and the potential problems that they might introduce when objectives are chosen, especially if specific performance characteristics are sacrificed in order to optimize another variable that might be more important in a particular application.

Uncorrected **astigmatism** can reduce image intensity, sharpness, and contrast, with increasing effects at greater distance from the optical axis (Figure 7.54). The geometry of an astigmatic image point may be defined by considering two orthogonal planes that are cross sections through the image wavefront. The planes (tangential and sagittal) in an astigmatic system may have different focal distances and exhibit separate radii for a perfectly symmetrical specimen point. This aberration results in symmetrical point-like features that are located off-axis being either radially or tangentially elongated in the image plane, depending upon focus. When the best focus is chosen at a compromise position between the two extremes, the resulting Airy disk is non-symmetrical, resulting in image degradation. Astigmatism can result from poorly centered elements in a low-quality or damaged objective, and is increased by other misalignments in the optical path of the microscope.

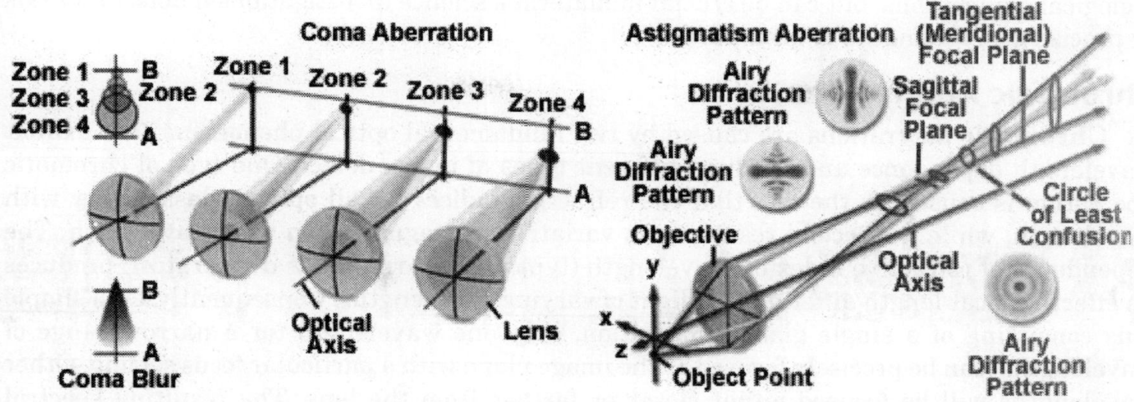

Fig. 7.54. Off-Axis Optical Aberrations in Microscope Objectives.

A lens property described as either **field curvature** or **flatness of field** is of major importance in many critical measuring and high-resolution imaging applications, such as in semiconductor inspection, although in most biological confocal investigations, it is not a serious limitation. Simple spherical lenses focus image points from different regions on a flat specimen onto a curved image surface, which reflects the shape of the lens surface. A flat image plane does not conform to the curved plane of focus, with the result that central and peripheral areas of the field cannot be simultaneously brought into sharp focus. More complex lens designs, consisting of multiple groups of lens elements, are necessary to correct this field curvature and to expand the size of the central zone of sharp focus. Objectives designated as **flatfield** or **plan** are corrected optically to produce a wide usable field having sharpness from center to edge, with minimal field curvature, at the intermediate image plane. The flatness of the field at the final image plane is also dependent upon intermediate optical elements, including the microscope eyepieces.

Modern objectives are capable of producing sharply focused image fields that are approximately twice the diameter of the usable field in early objectives. Specialized antireflection coatings are among the technological innovations that have allowed this achievement, although the increased optical complexity results in reduced light transmission and much greater cost. In confocal biological applications where thick sections or live cell preparations are imaged, the use of non-plan objectives may be tolerable because the section curvature produced by the objective is relatively insignificant given the uncertainty of other factors on the exact specimen shape. If other optical aberrations are well corrected, any potential distortion in three-dimensional data sets caused by slight field curvature can be corrected. In many cases, Nyquist sampling conditions may only require utilization of the central portion of the objective field.

Non-linearity of magnification from the center to the edges of the image field produces geometric **distortion** of specimen features, causing their true dimensional profiles to be skewed in the image. If present, this effect is readily observed when orthogonal sets of crossed lines are imaged, and rather than appearing straight over the entire image field, the lines are either curved outward or inward at regions away from the field center. These two types of distortion are commonly referred to as **barrel** and **pincushion** distortion, respectively. As is the case for field curvature, small amounts of geometric distortion are generally not highly significant in biological applications, but can be crucial in materials science investigations if defect analysis or precise measurements are intended.

Chromatic Aberrations

Chromatic aberrations are caused by two fundamental optical phenomena that exhibit wavelength dependence and produce different types of image defect. One type of chromatic aberration is caused by the fact that the refractive indices for all optical glasses vary with wavelength, while the second results from variation in magnification with wavelength. The dependence of refractive index on wavelength (typically referred to as **dispersion**) produces an effective focal-length difference for light of varying wavelengths. Consequently, for a simple lens consisting of a single glass composition, only one wavelength (or a narrow range of wavelengths) can be precisely focused at the image plane with a particular focus setting. Other wavelengths will be focused either closer or farther from the lens. The resulting spectral dispersion along the optical axis is termed **longitudinal chromatic aberration**, or alternatively, **axial chromatic aberration** (see Fig. 7.55). For a point-like source imaged on the optical axis, the gradation in refractive indices cause blue light to be focused nearest the lens, with longer wavelengths converging at focal points progressively farther from the lens. The aberration, if uncorrected, can be visualized as varying color fringes when the image is defocused to either side of the best visual focus.

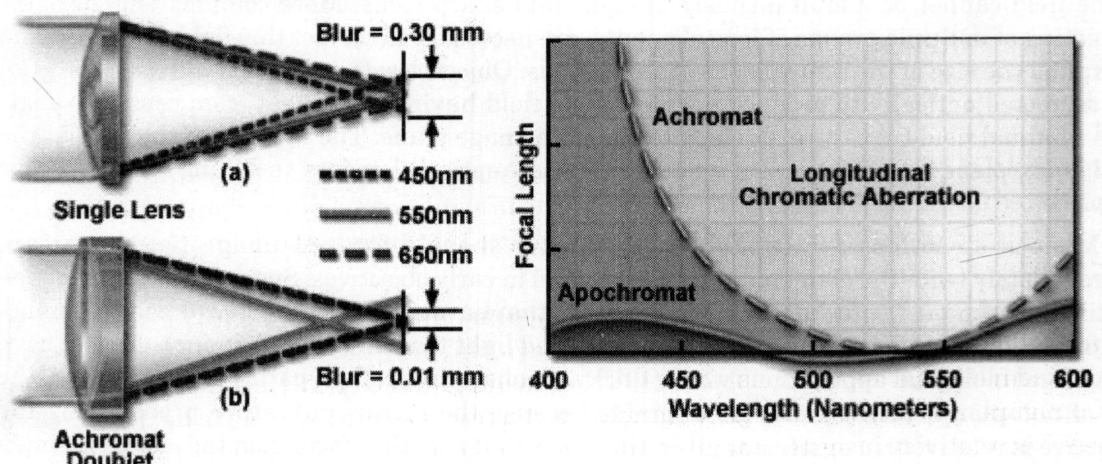

Fig. 7. 55. Chromatic Aberration in Confocal Microscopy.

Uncorrected longitudinal chromatic aberration in an objective used for confocal microscopy can have profound effects, particularly when imaging two or more fluorophores, which depend upon the capability to demonstrate co-localization of fluorescence emission at several wavelengths. When multiple fluorophores are employed, longitudinal chromatic aberration causes the various laser excitation beams to be focused at different points in the specimen, and similarly results in collection of the different emission wavelengths from non-coincident points. This aberration negates the possibility of establishing accurate positions of fluorophores in the **z**-direction for generation of specimen data in three dimensions.

In scanning confocal techniques, evaluation of the effect of chromatic aberrations on the integrity of images for a particular objective requires consideration of all factors that determine the signal energy that passes through the pinhole and is recorded by the detector. These factors include the excitation laser emission spectrum, the emission peak and bandwidth of each fluorophore, and the spectral sensitivity of the detector. The correction of longitudinal chromatic aberration is usually accomplished in the objective design by the combination of multiple lens elements having different optical characteristics, and the degree of correction forms part of the basis for the classification of objectives into different categories.

The change of lens focal length with wavelength, which produces the axial dispersion characterizing longitudinal chromatic aberration, is also responsible for the occurrence of **lateral chromatic aberration** (see Fig. 7.5). Because magnification is inversely proportional to focal length, the variation of focal length with wavelength produces a corresponding dependence of magnification on wavelength. If this aberration is not corrected in the objective, sharp edges in the image may be surrounded by red or blue color fringes. The magnification of the blue wavelength component from the signal may vary from that of the red component by approximately 1 to 2 per cent in uncorrected objectives. When present in a confocal scanning optical system, lateral chromatic aberration can result in a loss of signal at the pinhole, because of the possibility that emitted light will be imaged at a location that is either nearer to or farther from the optical axis than its true specimen location, depending upon the spectral composition of the light.

Because the two types of chromatic aberration are related, objectives that are highly corrected for longitudinal chromatic aberration generally exhibit minimal lateral chromatic aberration as well. Different approaches are taken by microscope manufacturers in correction of the various factors affecting optical performance, and careful matching of system components is essential to obtaining the maximum correction of aberrations. In confocal fluorescence microscopy, uncorrected lateral chromatic aberration results in similar problems to those of the longitudinal variety in that errors are introduced in mapping the locations of fluorophores emitting at different wavelengths.

For much of the history of optical microscope manufacture, a standard for objective design was followed that required the objective to form a real image at a specified fixed distance from the mounting surface of the objective, corresponding to the front focal plane of the eyepiece. Configurations based on the direct formation of a real intermediate image by the objective are referred to as **finite** systems, and the distance to the intermediate image is referred to as the **tube length** of the objective. Although at one time the various manufacturers standardized their finite optical systems on the same tube length and parfocal distance, different approaches have been employed to compensate for lens aberrations, and this factor must be considered if optical components from different manufacturers are used in combination. Finite systems may be designed to provide full aberration correction in the objective, or a known residual level of lateral chromatic aberration may be allowed in the objective-formed image, which is compensated by the eyepiece.

In the **infinity** optical system, light rays leaving the objective are not focused, but remain parallel until being converged at the intermediate image plane by a **tube** lens (occasionally referred to as a **telan** lens). Most major microscope manufacturers have developed infinity-corrected optical systems, which have the intrinsic benefit that light rays from the objective, focused to infinity, are relatively insensitive to additional optical components placed in the "infinity space" between the objective and tube lens. Although the tube lens can perform some correction of residual aberrations, there are advantages in versatility of designs that provide full correction within the objective, with the tube lens being optically neutral. Presented in Figure105 are the infinity light pathways for typical upright and inverted fluorescence microscopes. The double-headed white arrow in each microscope diagram indicates the parallel light path between the objective rear aperture and the tube lens.

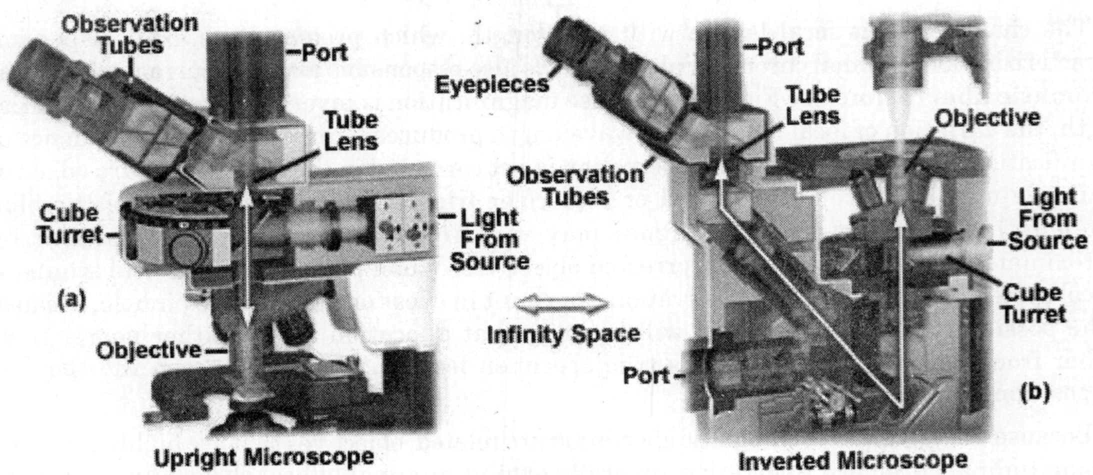

Fig. 7.56. Parallel Infinity Space in Upright and Inverted Fluorescence Microscopes.

Different manufacturers adhere to different design specifications in their infinity systems, including tube lens focal length and length of the infinity space. Infinity-corrected optical systems enable the objective to be moved for focusing instead of the microscope stage, a distinct advantage in applications requiring delicate manipulation of the specimen during observation. However, the principal practical benefit of the infinity space is that auxiliary optical components, such as polarized light analyzers, filters, and differential interference contrast (**DIC**) prisms, can be added without major concern for their refractive properties or thickness. As long as they have plane parallel surfaces, the added elements have negligible effect on image quality. In contrast, optical components cannot be placed into the beam path of a finitely corrected system, which is converging between the objective and intermediate image plane, without introducing image shift and other aberrations that vary with the thickness and refractive index of the added elements.

Optical Correction in Microscope Objectives

Regardless of whether a finite or infinity-corrected optical system is employed, the combined performance criteria of the objectives designed for the system must be taken into account in determining which objectives will meet the requirements of a particular imaging technique. Objectives are typically grouped into performance categories based upon their degree of optical correction. As discussed, the relative importance of various aberrations and the performance criteria they affect depends upon the manner in which the objective is used and the essential requirements of the application. The most general grouping of objectives is typically based on their degree of chromatic aberration correction, although the traditional objective categories have become much less distinct recently, due to significant technological advances in optical design and manufacturing. Much of the descriptive nomenclature is still valid and widely used, and an understanding of the terms employed and how they pertain to confocal applications is important when the various objectives available are being evaluated.

Chromatic aberrations are a result of dispersion in the optical glass used to fabricate a lens, and are typically corrected by utilizing a combination of lens elements having different dispersive properties. Traditionally, objectives having the minimal correction for chromatic aberration are categorized as **achromats** (see Figure 7.57), and are usually constructed using glass having "normal" dispersion. Glasses of this type exhibit a nearly linear decrease of refractive index with increasing wavelength, and include crown and flint glasses. Crown glass usually exhibits low refractive index and low dispersion, in contrast to flint glass, which typically has high refractive index and high dispersion. In older achromatic objectives, two or more lens elements of these glass types were combined to provide chromatic aberration correction that brings red and blue light to a common focus, and also corrects spherical aberration for green light. Modern achromatic objectives typically have additional correction for spherical aberration, and significant correction for field curvature. If the objective is corrected to extend flatness of field over the entire image field, it is given the designation **plan achromat**. In general, achromatic objectives are suitable for conventional brightfield visual observation, and for photomicrography and digital imaging with the additional (**plan**) correction for field curvature.

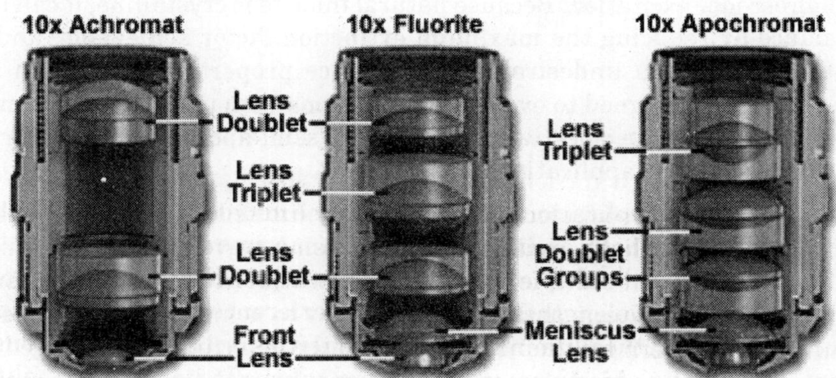

Fig. 7.57. Common Objective Optical Correction Factors.

In order for the objective to achieve improved chromatic aberration correction, glass types having abnormal dispersion over a portion of the spectrum are required. In glasses that exhibit non-linear change of refractive index with wavelength in the red or blue spectral region, the chromatic aberrations can be offset to allow simultaneous focus of more than two wavelengths. Crystalline fluorite was one of the first materials found to provide suitable optical properties in reducing the uncorrected secondary spectrum responsible for the green or purple fringes bordering sharp edges, which characterize images acquired with achromats. In combination with various optical glasses, fluorite elements provide the capability to correct chromatic lens aberrations for three wavelengths (colours) and spherical aberration for two. Additionally, such highly corrected objectives have improved transmission characteristics in the ultraviolet spectral region.

More recent technological developments have provided new optical glass formulations and lens shaping capabilities that produce dispersion properties similar to those of fluorite elements, and most manufacturers produce a line of **fluor** objectives that have optical corrections approaching those of the highest category (Figure 7.57). These objectives are sometimes referred to as **semi-apochromats**, and may or may not contain fluorite elements, but because of their optical characteristics, are nevertheless designated by different manufacturers with names such as **Fl**, **PlanFl**, **Fluor**, **Plan Fluor**, **Neofluor**, and **Plan Neofluor**. Current objectives in this category are available in many configurations, including models for use with multiple immersion media, and are suitable for use in brightfield, fluorescence, phase contrast, polarized light, and differential interference contrast, as well as some confocal and multiphoton applications.

Objectives classified as **apochromats** are generally those that have the highest level of correction for chromatic and spherical aberrations (see Figure 7.57). These objectives typically have the highest available numerical apertures for a given magnification, and both chromatic and spherical aberrations are corrected for at least three wavelengths. With nearly complete correction of aberrations, apochromatic objectives are generally suitable for any microscopy technique, although all of the specific performance requirements for the technique being employed must be considered. In spite of the exceptional degree of optical correction exhibited by apochromats, because they utilize lens elements of abnormal dispersion (usually including fluorite), they may not be the best choice for certain polarizing microscopy applications or for ultraviolet fluorescence excitation. Because natural fluorite is crystalline, it can limit polarized light performance by reducing the maximum extinction factor achievable, and may contain organic contaminants with undesirable fluorescence properties. Unless an apochromatic objective is specifically designed to overcome limitations with polarized light and in the near-ultraviolet spectral region, an objective in the fluorite (semi-apochromat) category may provide better performance in some applications.

Confocal fluorescence applications can be severely limited if the objective aberrations are not similarly corrected for both excitation and emission wavelengths, a requirement made more difficult to satisfy when multiple fluorophores are utilized or when the difference between excitation and emission wavelengths is large. In order to achieve maximum detected photon energy, parfocality between the illuminating spot and the detected region imaged at the pinhole must be maintained. Many objectives, even in the apochromat class, do not provide adequate correction for fluorescence techniques combining ultraviolet excitation with visible-region emission. Additional optical components may be employed at the ultraviolet laser source to compensate for the failure of the objective to accommodate the wide range of wavelengths, although this adds expense and considerable operational complexity to the confocal system.

High-performance water immersion objectives have been recently introduced by most major manufacturers, and many of these objectives are specifically designed to satisfy requirements imposed by current rapidly developing research areas. A primary goal is achieving optimum performance in confocal fluorescence microscopy of biological specimens, including live cells and tissue supported in physiological media. It is essential to recognize that regardless of the level of aberration correction incorporated into an objective's design, additional aberrations can be introduced into the optical path outside the objective by violating the design's operating requirements, and these can degrade the performance of the best components.

The newer water immersion objectives (see Fig. 7.58) avoid one of the most common problems of this type, which is the refractive index mismatch between immersion and specimen media. They also provide high resolution and improved light collection through large-numerical aperture design, and have relatively long working distance. Additionally, specific attention is given to high transmission of ultraviolet wavelengths, and to extending the optical correction from the ultraviolet to the visible emission region for common fluorophores, in order to maintain confocal conditions for excitation and signal collection. Olympus, for example, offers a range of specialized **W LSM** water immersion high numerical aperture apochromats (for confocal microscopy; see Fig. 7.51), which are corrected for a large flat field, and have extended apochromatic correction from approximately 340 to 650 nanometers.

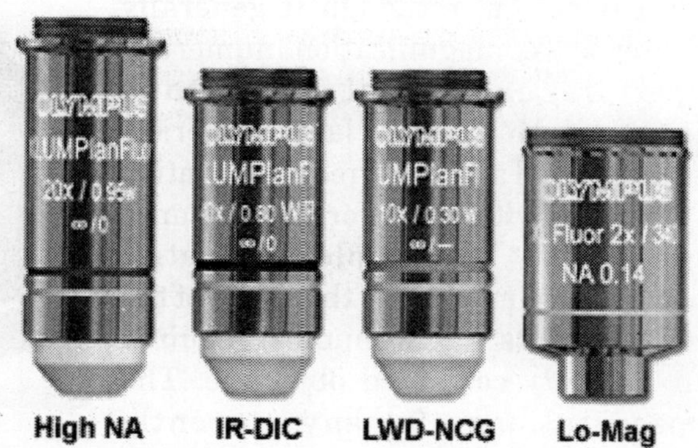

Fig. 7.58. Water Immersion Objectives.

Objectives designed for direct immersion or water dipping, which feature specialized ceramic or polymeric nosecones, are available for live cell and physiological studies (Fig. 7.51). These investigations often require access to the specimen for measurements or other manipulations, and cannot be done with a cover glass in place. Generally water dipping objectives are in the fluorite category, have high transmission in both the ultraviolet and infrared spectral regions, and are designed with long working distance specifications and narrow profile of the insertion (dipping) portion at the nose of the objective. The smaller angular nose on the objective is intended to allow maximum access to the specimen for attachment of microelectrodes or to perform other operations during observation. Long working distances, which may be several millimeters with some objectives of this style, also offer greater access to the specimen. The immersion portion of many dipping objectives is constructed from ceramic or another inert insulating material for electrical isolation and chemical resistance. Their combination of characteristics makes water dipping objectives suitable for live specimen confocal, multiphoton, and many other imaging methodologies.

In addition to magnification, numerical aperture, and degree of optical correction, an objective's **working distance** has particular significance in confocal microscopy and other digital microscopy techniques used for acquiring three-dimensional specimen information. The working distance, for an objective requiring a cover glass, is defined as the distance between the top surface of the cover glass and the front lens element when a specimen plane in contact with the cover glass is in focus (Figure 7.59(b)). This distance is important when confocal imaging is performed at various depths in the specimen because it determines the maximum focal penetration depth that is attainable before the objective contacts the top of the cover glass. The working distance, therefore, limits the range along the **z**-axis from which specimen data can be collected.

As an example, an objective with 0.20-millimeter (200 micrometers) working distance can be focused to a maximum depth of 200 micrometers below the cover glass before the objective contacts the cover glass upper surface. Within limits, the working distance can be varied by optical design, although it generally decreases as magnification, numerical aperture, and optical correction is increased. Maintaining large numerical aperture and a high degree of aberration correction with long working distance is limited by simple geometric constraints imposed by the shape of the lens elements and the number required in a highly corrected objective. The specifications of many currently available objectives represent remarkable achievements in performance realized by the availability of improved optical glasses, lens coatings, and computer design capabilities.

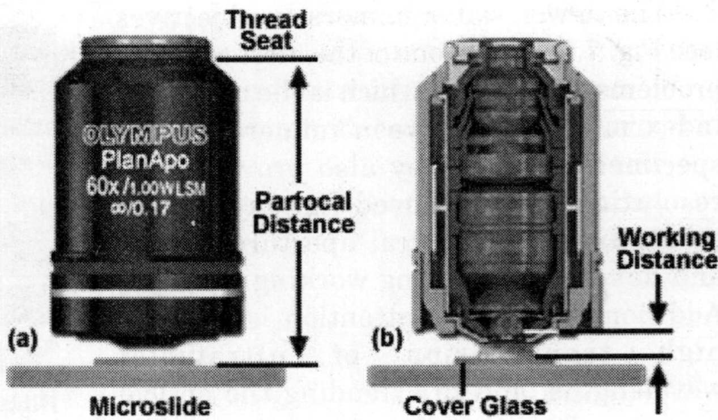

Fig. 7.59. Objective Parfocal and Working Distances.

The fact that signal intensity is inversely proportional to the square of magnification has implications in the choice of optimum objectives for confocal microscopy, in some cases altering the relative importance of various performance factors. Some researchers have concluded that for imaging subcellular components of thick tissue sections, it is usually advantageous to select moderate-magnification high numerical aperture immersion objectives from the semi-apochromat category in order to obtain maximum light collection. When utilizing scanning confocal systems with an electronic zoom feature, it is seldom necessary to employ 100x optical magnification, and the highly corrected (and high numerical aperture) 40x and 60x objectives may be more suitable in techniques benefiting from the additional light collection.

Many popular techniques, such as immunofluorescence imaging, inherently suffer from such low light levels that the transmission properties of the optical system are of crucial importance. Only a few photons may be available for the detection of minute specimen details, and as a consequence a relatively minor difference in transmission per centage at the wavelength of fluorophore excitation or emission can be critical in determining the detectability of a labeled feature. In some situations, the objective's transmission properties may have greater significance than other specifications such as flatness of field and aberration correction, which require additional lens elements and coatings that can increase light loss in a critical wavelength band. Apochromatic correction may compromise other objective requirements, as discussed previously regarding polarized light performance, and reduced light transmission in the ultraviolet or near-ultraviolet range has traditionally been one of the costs of full aberration correction.

The goal of improving transmission properties across a wider range of wavelengths while still achieving a high level of aberration correction has been considered crucial in confocal fluorescence applications, and many of the recent high performance objectives, including those for water immersion, have been optimized for high transmission in the ultraviolet. It is important to consider the transmission properties of an objective at the wavelengths for the technique being utilized, and such data is usually available from the objective manufacturer. The specific spectral transmission properties are determined by the optical glass composition, the cements used between compound lens elements, and the coatings applied to the lens surfaces. As the sophistication of lens design has increased, the importance of lens coatings to reduce internal reflections has become increasingly importance. Earlier simple single-layer coatings substantially improved overall transmission, and the continuing refinement of proprietary materials for multilayer coatings has further enabled lens designers to expand the transmission of objectives in desired wavelength bands, even though the number of lens elements employed and overall complexity of the objectives have increased significantly.

The rapidly developing arena of specialized techniques being applied to problems in cellular and molecular biology, as well as in other fields, has resulted in changes to the demands placed on optical imaging systems, particularly with regard to the microscope objective. The dramatic increase in utilization of confocal scanning techniques for three-dimensional investigations of biological specimens is a major factor affecting the products being commercially developed, although methods such as laser trapping, multiphoton excitation, fluorescence resonance energy transfer (**FRET**), fluorescence ***in situ*** hybridization (**FISH**), and infrared differential interference contrast have also introduced new requirements. A number of the parameters for specialized techniques have made it apparent that certain traditional objective performance criteria may be able to be at least partially sacrificed in order to improve other more critical specifications. Dramatic advancements in computer-aided lens design and optical glass formulations, in addition to those in antireflection coating technology, have resulted in the introduction of improved optics that are able to meet many of the new requirements, with few or no trade-offs in conventional wide field microscopy. Examples of these are the high performance water immersion objectives with higher numerical aperture, increased working distance, spherical aberration correction collars, and objectives with enhanced ultraviolet and infrared transmission.

Many modern objectives are suitable for use in confocal microscopy when they are utilized in accordance with their design specifications. For many applications, objectives with less than full apochromatic correction, such as fluorite-class semi-apochromats, are an ideal compromise as an all purpose objective that can serve in many confocal techniques. High-resolution examination and accurate three-dimensional imaging of thick aqueous specimens with the laser scanning confocal microscope is particularly demanding and requires an objective satisfying all of the following criteria: high numerical aperture for efficient fluorescence collection, long working distance to allow maximum penetration depth, flat field for accurate three-dimensional reconstruction, low axial chromatic aberration for parfocality of multiple fluorophores, and low lateral chromatic aberration to enable precise registration of multiple fluorescence images, in addition to high transmission at the excitation and emission wavelengths.

Confocal Microscope Scanning Systems

Confocal imaging relies upon the sequential collection of light from spatially filtered individual specimen points, followed by electronic signal processing and ultimately, the visual display as corresponding image points. The point-by-point signal collection process requires a mechanism for scanning the focused illuminating beam through the specimen volume under observation. Three principal scanning variations are commonly employed to produce confocal microscope images. Fundamentally equivalent confocal operation can be achieved by employing a laterally translating specimen stage coupled to a stationary illuminating light beam (**stage scanning**), a scanned light beam with a stationary stage (**beam scanning**), or by maintaining both the stage and light source stationary while scanning the specimen with an array of light points transmitted through apertures in a spinning **Nipkow disk** (see Figures 86 and 109). Each technique has performance features that make it advantageous for specific confocal applications, but that limit the usefulness in others.The stage-scanning and beam-scanning configurations are **single-beam** methods, while the spinning disk approach is a **multi-beam** scanning technique. Systems employing the Nipkow disk-scanning concept have typically utilized noncoherent broad-spectrum light sources (such as arc-discharge lamps) for illumination rather than lasers, and the overall lack of brightness has severely limited their use in fluorescence applications. More recently, microlens arrays and sophisticated disk design enhancements, coupled to laser illumination, have expanded the potential applications for spinning disk confocal microscopes. Nipkow disk systems may be designed in either **tandem scanning** or **monoscanning** variations. In the former, the illumination and detection beams follow tandem pathways through separate sets of identical apertures located on diametrically opposite sides of the disk. The monoscanning system performs illumination and detection simultaneously through each rotating aperture on the disk, maintaining the coincidence that the two light paths exhibit while passing through the objective.An additional method of single-beam scanning has experienced limited application in specialized reflected light microscopes, primarily for integrated circuit inspection. The objective itself may be scanned over a stationary specimen utilizing a stationary light source in a **scanned-lens** system. This configuration has similar optical advantages to stage scanning, but allows the stationary specimen to be fitted with measuring probes, or otherwise manipulated. The configuration is not amenable to rapid scanning when a relatively massive conventional objective is employed, and is therefore not widely utilized.The modern confocal microscope is an integrated electronic system, most commonly based on a widefield epi-fluorescence instrument, with the addition of multiple laser illumination sources, a scan head containing electronic and optical components, a computer and monitor for image display, and associated software for control of signal acquisition, processing, and image analysis. In the basic confocal optical configuration, the objective forms an image of both the source and detector pinholes on the specimen plane. By positioning the pinholes on the microscope optical axis in conjugate focal planes, their images overlap within the specimen focal plane. Although fluorophores outside the plane of focus are excited, detection is limited to emission occurring near the focal plane by the detector pinhole aperture, which rejects out-of-focus light. The single-beam laser scanning confocal microscope functions in this **point-scanning** mode as a sampling device and does not form an optical (real) image. To enable image formation, the sampling spot must be moved through the specimen and the resulting signal collected and stored. The scan head controls the generation of photon signal required to construct the confocal image. The components of a typical commercial scan head are illustrated in Figure 1, and generally include one or more laser inputs, fluorescence filter sets, a raster scanning mechanism, one or more pinhole apertures, and detectors (usually photomultiplier tubes, **PMTs**) for multiple fluorescence wavelength detection.In order to extend the confocal point-sampling principle to allow generation of an extended specimen image field, the point focus in the specimen is scanned in a raster pattern similar to that employed to create the image on a television screen (and in other video applications. This mechanism requires a fast horizontal scan (the line scan) in conjunction with a slower vertical scan, or frame scan, which offsets the scanning line to sequential positions from top to bottom of the frame. During the history of confocal microscope development, a number of different techniques have been employed to implement point scanning, and several have been refined into current commercial versions. In single-beam laser scanning instruments, a typical raster scanning mechanism utilizes two high-speed oscillating mirrors driven by galvanometer motors, which pivot on mutually perpendicular axes. Coordination of the two mirrors, one scanning along the **x**-axis and the other on the **y**-axis, produces the rectilinear raster scan. The scanning speed of the mirrors is negligible compared to the speed of light, and consequently the emitted fluorescence can be collected by the objective and returned, or **descanned**, along the original illumination path to its conjugate focal plane at the detector pinhole. Variation in the signal intensity illuminating the detector aperture corresponds to variations in emission at different points in the specimen as the exciting beam is scanned.

Several characteristics of an ideal confocal scanning system, required to provide optimum imaging performance, are extremely difficult to achieve in practice. Nearly all scanning system configurations present some shortcomings in operation, and various optical and electronic design modifications have been introduced in an attempt to correct the deficiencies. Several methods of implementing point scanning reduce sensitivity or involve severe compromises in flexibility or image quality, and are not currently used in commercially produced systems. One of the most important requirements of the scanning system design is that the objective pupil (rear focal aperture) is completely filled with light during the entire scanning cycle in order to prevent illumination fall-off at the scan extremes. This is best accomplished by minimizing movement of the beam at the aperture with a scan design that pivots the beam at a stationary point that is conjugate with the objective rear aperture. Maintaining a stationary pivot point when the beam is rocked during scanning is technically challenging, and some systems compensate for a small amount of beam movement by overfilling the aperture through increased beam expansion. This has the disadvantage of wasting light and reducing the photon efficiency of the system.

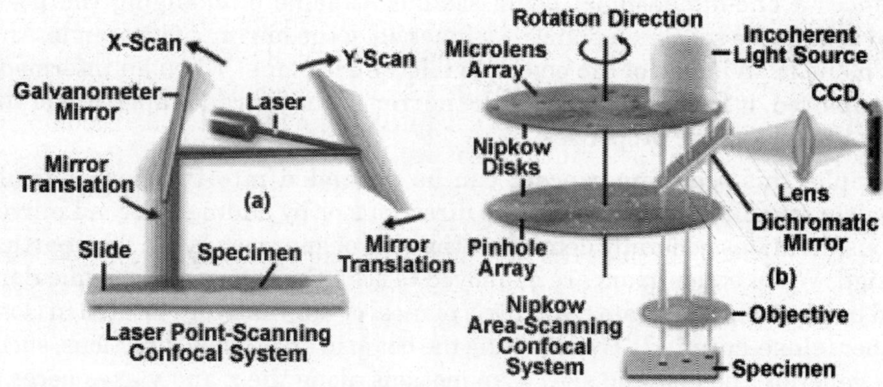

Fig. 7.60. Point and Area-Scanning Confocal System Configurations.

An additional desirable property of the confocal scanning mechanism is to scan at the highest frame rate possible in order to provide the flexibility to adjust various scanning modes to match the imaging application. This requires minimal inertia of the moving components that produce the beam scanning, as well as minimizing the system **dead time**, (or interval between each scanning cycle) in which the beam is not scanning the specimen. The proportion of each full-frame scanning interval that is utilized to actually scan the specimen is referred to as the **duty cycle** of the system. Minimizing non-productive scanning of the beam is not only essential to achieving a high frame rate, but in some instrument designs, reduces the unnecessary specimen photon damage that results from a poor duty cycle specification.

The capability to freely rotate the scan raster around the optical axis is a very important feature for confocal imaging as a means to optimize the scan direction with respect to specimen shape or other characteristics. When imaging elongated features such as fiber bundles, orientation of the fast scan direction parallel to the long axis of the feature greatly enhances the time resolution of the specimen signal. Furthermore, the ability to rotate the raster allows specimen features to be oriented in a manner that utilizes the image field most efficiently. Scanning arrangements that do not permit rotation of the raster direction can seriously limit the practicality of the system, unless the specimen itself can be easily rotated, a much more problematic operation that is not usually feasible.

The optical arrangement required to produce a linear motion of the illumination spot in the specimen is derived from consideration of the geometrical optics of the microscope, including the fact that the objective is telecentrically corrected (a telecentric lens system positions the entrance and exit pupils at infinity). To realize the full optical correction of the objective, the image and specimen planes must remain at fixed distances from the objective, and the locations of the conjugate image planes and conjugate telecentric planes are therefore known. A critical optical property is that all light beams intersect a telecentric plane at an angle that is a function of the position of the source point in the specimen plane. Since a flat mirror is capable of changing the angle of propagation of a light beam, the placement of a mirror at a conjugate telecentric plane on the optical axis provides a mechanism by which a change in beam angle will produce a linear motion of the focal point in the specimen. Therefore, in the simplest case, the positioning of a mirror with its pivot point at the center of the objective conjugate telecentric plane produces a one-dimensional beam scanner capable of changing the position of the illuminated spot in the specimen plane as a function of the mirror's pivot angle. Any conjugate telecentric plane is an image of the objective telecentric plane. When an intermediate optical system is employed, it forms an image of the mirror in the entrance aperture of the objective, maintaining the telecentric properties.

In principle, this scanning concept can be expanded into two perpendicular axes by simultaneously scanning the mirror in two directions or by adding a second mirror, although practical considerations generally determine the type of approach taken for a particular overall system design. When two mirrors are employed to scan the beam in perpendicular directions, they should be placed in conjugate telecentric planes, or alternatively, located in close proximity to one another (**close-coupled**). By deflecting the beam in orthogonal directions, such a scanning system can generate the fast and slow scan motions along the **x** and **y** axes necessary to form a complete two-dimensional image.

Various arrangements of scanning system components are possible provided that the primary requirements are satisfied. To ensure diffraction-limited performance of the optical system, the objective rear focal plane (entrance aperture) must constantly be uniformly filled by a planar wave during scanning. Because the physical diameter of this aperture varies with properties of the objective, all other components, including illumination pinholes, must be matched to the objective(s) in use. Conjugate telecentric planes can be produced in required locations by the addition of auxiliary optics, and if this approach is taken, the properties of these must also be carefully considered in regard to their compatibility with the objectives chosen for use with the system. The beam properties of illuminating lasers, in particular the diameter of the Gaussian beam profile, are significant factors in adjustment of pinhole diameter and other variables related to illumination of the objective entrance aperture.

In the simplest beam-scanning confocal configuration, a scan mirror is located in the rear focal plane of a scan lens, which is conjugate with the rear focal plane of the objective. Illustrated in Figure 110(a) is a single-mirror arrangement, which includes the tube lens required by an infinity-corrected objective. Scanning on one axis is easily accomplished with this configuration. The theoretically ideal means of achieving **x-y** scanning is to scan a single mirror on both axes simultaneously (referred to as **cardanic** scanning). Two scanning mirrors are more commonly employed, and two such possible configurations are illustrated in Figs. 7.61(b) and 7.61(c). If the mirrors are close-coupled (Figure 110(b)), the system can function satisfactorily without requiring additional intervening optics. With greater separation distance between the scanning mirrors (Fig. 7.61(c)), a multiple-lens telecentric relay system must be utilized to optimize the optical performance.

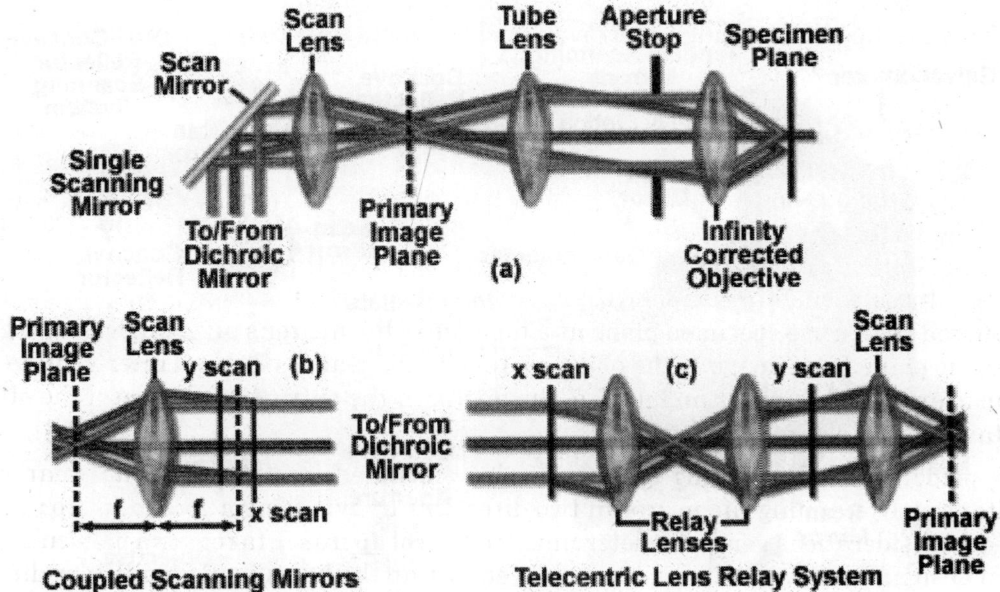

Fig. 7.61. Confocal Microscopy Scanning System Configurations.

General descriptions for several possible configurations of **x-y** scanning systems, which have been incorporated into commercial confocal microscope systems, are presented in the following discussion. Note, however, that not every arrangement can be implemented with all optical systems, and additional variations and refinements are continuously being developed by the manufacturers. Current systems offered in the market may not purely conform to any of the fundamental configurations described.

One of the most common scanning arrangements combines two independent scan mirrors with a relay optical system, as illustrated in Figure 7.62(a). The simultaneous movement of the mirrors on perpendicular axes produces the raster scan on the specimen. The intermediate optics form an image of one mirror on the axis of the second, and consequently the placement of one mirror in a telecentric plane ensures that the other will occupy a conjugate telecentric plane. The configuration meets the optical requirements perfectly, although the intermediate optics must be extremely well corrected. As is the case for objectives, the highest level of aberration correction requires increased optical complexity, resulting in reduced luminous efficiency for the system. The scanning mirrors are typically oscillated electromechanically by a moving-coil mechanism similar to that of a galvanometer, and are commonly referred to as **galvo** mirrors. A novel modification of this arrangement is employed by Bio-Rad in that company's confocal and multiphoton instruments: two off-axis concave reflectors are utilized to image the first scanning mirror onto the second [Fig. 7.62(b)]. This allows the beam to be scanned by the first mirror over the surface of the concave mirrors, then directed to a single point on the second mirror. The arrangement has the advantage that the beam scans through an angular range about a fixed point on the second mirror, avoiding any beam shift. The scan mirrors can be small relative to the fixed concave mirrors, minimizing their inertia, and allowing rapid oscillation for fast frame rates, as well as enabling scan rotation.

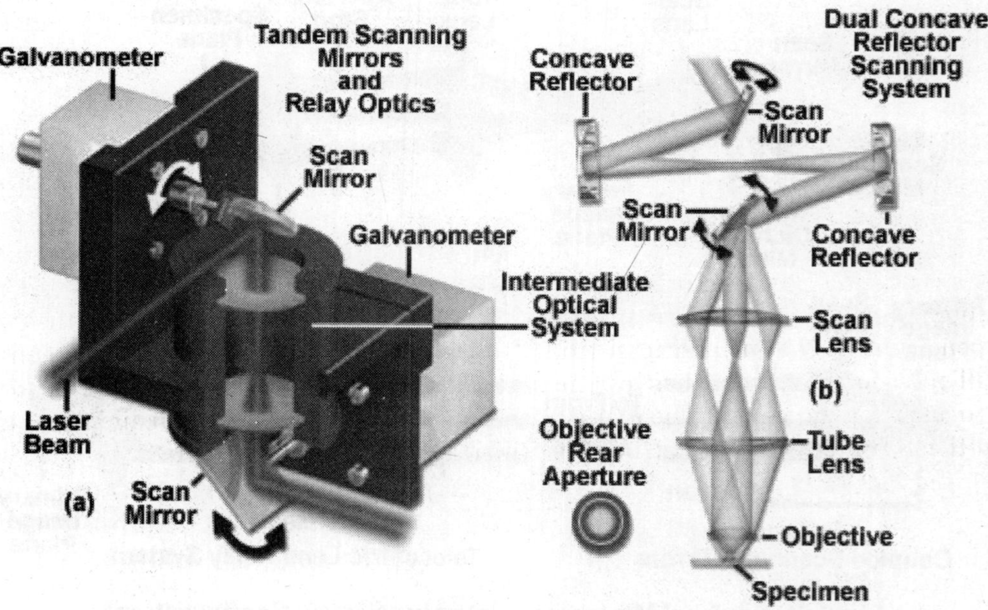

Fig. 7.62. Confocal Microscope Scanning System Configurations.

The two-mirror configuration can be utilized without the intervening relay optics by positioning the scanning mirrors as closely as possible, with a shared telecentric plane located at the midpoint between them (Fig. 7.63(a)). This system design has the advantage of optical simplicity and compact size, and has been implemented in instruments developed by Zeiss and other manufacturers. The design achieves only approximate telecentricity, although the closer the mirrors are to each other, the smaller the error. The deviation from ideal performance varies with the characteristics of the microscope intermediate optical configuration, and is usually only significant at low zoom settings, for which large scan angles are required.

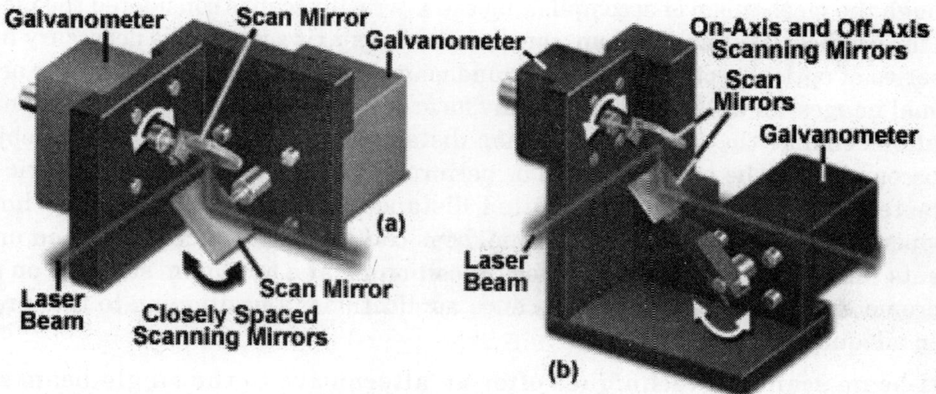

Fig. 7.63. Confocal Microscope Scanning System Configurations.

An improvement of the close-coupled mirror arrangement, which has been employed by at least one instrument manufacturer, is to position one scan mirror so that its rotational axis is slightly different from the illumination beam axis (Figure 112(b)). This has the effect of shifting the mirror as it rotates, and improves the accuracy with which the mirrors approach the conjugate telecentric plane. The shifting movement of the second mirror is intended to substitute for the function of relay optics, which are not utilized in the close-coupled design. A disadvantage of this design is that successive scan lines in the image frame have slightly different optical path lengths, making the approach unsuitable for use with many microscope optical systems.

Scanning a single mirror on two orthogonal axes is the simplest method of generating a two-dimensional scan raster (see Fig. 7.64). A cardanic scanning system of this type has been implemented by Leica in at least one confocal microscope model. The basic design used is illustrated in Fig. 7.64(a), and functions by incorporating the scan mirror in a small galvanometer head, which itself is mounted on a second larger scan device whose rotational axis passes through the mirror center perpendicular to the galvanometer axis. The fast line scan on the **x**-axis is accomplished by the small galvanometer, which has low inertia and can respond quickly, while the larger scanner is responsible for the **y**-scan. The single-mirror arrangement benefits from simplicity and the fact that the mirror is easily located in a telecentric plane of the optical system.

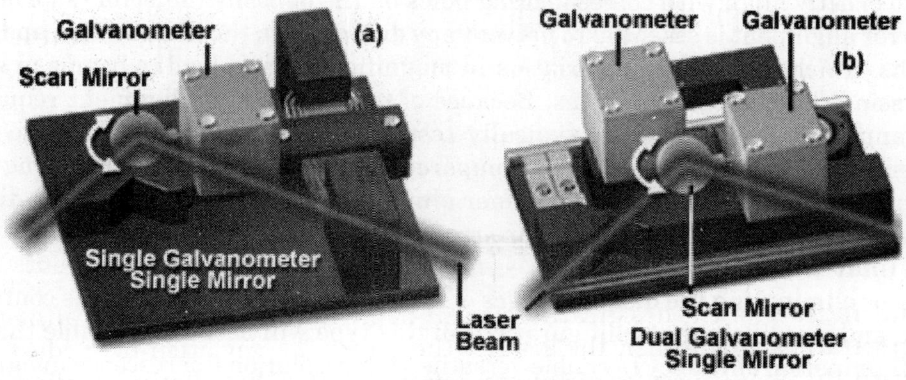

Fig. 7.64. Confocal Microscope Scanning System Configurations.

Although the mechanism of accomplishing the **x-y** raster scan is considered the most crucial aspect of the confocal scanning system, some method of **z**-axis scanning is necessary in order to acquire series of optical sections for three-dimensional imaging, and to collect **x-z** or **y-z** two-dimensional images, as well as to perform any form of free-line **z**-scanning. Typical microscope configurations change the objective-specimen distance by translating either the objective or the microscope stage. The movement can be performed precisely with a piezoelectric driver or galvanometer device, albeit over a limited distance range. More commonly, however, a microstepper motor is used to drive the fine-focus control of the microscope, and on current instruments the steppers are capable of focus positioning at a minimum step size on the order of 10 micrometers. For biological fluorescence applications, **z**-positioning to this precision is more than adequate.

Multi-beam scanning techniques offer an alternative to the single-beam scanning configurations, although low illumination efficiency has previously limited their use in high-resolution fluorescence applications. In either the tandem or monoscanning variation, rotating-disk scanners place hundreds of holes, which function as illumination and detection pinholes, in the microscope intermediate image plane. The configuration of a typical Nipkow disk scanning system is presented in Figure 7.61(b). The holes in the disk are arranged so that a large number of beams uniformly scan the image field as the disk rotates, completely covering the specimen at a much higher rate than that of single-beam scanners. Because disk-scanning microscopes form a real image, a CCD camera may be located directly in the image plane, collecting emitted signal at much greater quantum efficiency than that exhibited by photomultiplier tubes. In spite of this advantage, which allows real-time focusing and imaging of dynamic processes, several shortcomings have limited the practical utility of disk-scanning confocal systems. One of the most serious is the typical reliance on conventional broad-spectrum light sources, and the extreme light loss that occurs at the disk. Advances in design of disk systems and the application of laser sources have helped overcome some of the efficiency problems.

Because each disk has holes of fixed size, the pinhole diameter cannot be matched to the specific objective being used, and consequently the choice of objectives that will perform optimally with a given disk is limited. Furthermore, it is not possible to optimize the diameter of source and detector pinholes independently. Unique optical alignment difficulties exist with the tandem-scanning systems, which require the light rays returning from illumination points on the specimen to exactly align with corresponding holes on the opposite (detector) side of the disk. Precise mirror alignment is essential to prevent any difference in the illumination and detection path lengths, which could cause a variation in magnification and lead to beams in some disk regions missing the detection pinholes. Because of the demanding alignment requirements, tandem-scanning disk systems are typically restricted to using relatively large pinholes, compromising optical section thickness compared to single-beam laser scanning systems. Monoscanning variations of the disk scanner eliminate many of the alignment difficulties, although somewhat elaborate measures are required to reduce reflections from the disk that otherwise limit the detectable signal. Numerous technological improvements aimed at minimizing or eliminating the disadvantages of disk-scanning confocal systems continue to be introduced, and it is likely that configurations of this type will become available that provide sufficient illumination efficiency to enable versatile high-resolution fluorescence imaging within a practical time frame.

One technique for improving upon the lack of brightness that characterized early Nipkow disk systems is to utilize microlenses to intensify the source light. Older systems transmitted only approximately one per cent of the illumination incident on the disk, and necessitated the use of either cooled CCD or intensified video cameras to compensate for the low signal level. Newer disk scanning microscope designs incorporate a second disk containing thousands of microlenses, which spins in alignment with the Nipkow disk, and amplifies the light passing in both directions through the Nipkow disk apertures. Some advanced models combine dual spinning disks and microlens technology with laser illumination and cooled CCD detectors to allow fast data acquisition at normal video rates or faster.

Various additional scanner modifications have either been proposed or implemented in the effort to improve some aspect of the practical performance of confocal instruments. Every aspect of confocal fluorescence microscopy is fundamentally related to efficiency, and the inherent limitations of serial data collection. The efficiency with which useful signal can be acquired defines the balance that must be reached between image contrast and photodamage to the specimen, and also controls the required compromise in serial data collection between spatial resolution of the sampling scan, the signal-to-noise ratio, and the rate of image acquisition. The issue of efficiency has particular implications in attempts to perform video-rate single-beam imaging.

There is a strong interest in fast-scanning single beam instruments that can provide imaging at video rates in order to follow dynamic processes in living cells. One approach to such high image frame rates is to utilize rapidly oscillating resonant mirror scanners. Rotating polygon mirror scanners can achieve very high scan speeds, and are used in many optical devices, but do not provide the illumination and detection precision necessary for implementation in high-resolution microscopy. Various configurations combining scanning mirrors with acousto-optic deflectors (**AODs**) have been explored. In some arrangements, an AOD provides very rapid scanning on one axis, and a mirror scanner controls the slower axis. This approach is acceptable in some applications, but is problematic in confocal fluorescence imaging because it does not allow the longer-wavelength fluorescence emission to be descanned back through the acousto-optic modulator, which is wavelength-specific. The partially descanned signal, still oscillating on one axis, can be passed to a photomultiplier through a slit aperture, or imaged onto a linear-array CCD detector. Although the resulting images are confocal on only one axis, the characteristics are acceptable for some applications.

For high-resolution fluorescence imaging, the most refined and versatile scanning technology currently available is some variation that utilizes galvanometer scanners, and most major microscope manufacturers produce at least one confocal instrument employing this methodology. Because of the importance of photon efficiency, the relative simplicity of single beam scanning techniques give them definite advantages over disk scanners in most fluorescence applications. They are compatible with a wide variety of conventional microscope optical systems and video equipment, they conform well to general microscopy principles, and the flexibility in pinhole adjustment allows optimization to specific optical and specimen variables.Improvements that have been incorporated over several instrument generations have now culminated in a relatively stable technology with the versatility to accommodate new lasers and fluorescent dyes, which are being continuously developed. With the exception of resonant mirror systems, the galvanometer scanners are not capable of true video-rate imaging, although the implementation of bidirectional high-speed scanning can provide frame rates on the order of 0.25 seconds at reasonable pixel resolution. The benefit of single-beam scanning at video rates is questionable in any case since the photon limitations of acquiring fluorescence data at high magnification makes it difficult to achieve an adequate signal-to-noise ratio, regardless of how the beam scanning is accomplished. If the fluorescence signal is so low that several video frames must be integrated to achieve high-quality images, then there is no advantage in temporal resolution over imaging at slower scan rates, and many experimental approaches are now providing valuable data pertaining to dynamic events, utilizing current state-of-the-art galvanometer scanning systems.

Conclusions

Among the major differences between multiphoton and confocal laser fluorescence microscopy is the type of laser utilized in these often complementary techniques. Lasers for multiphoton microscopy are considerably more expensive and difficult to operate than the small air-cooled lasers employed in confocal microscopy. Development of user-friendly turnkey laser sources for multiphoton microscopy is a necessity if the technique is destined to enjoy widespread acceptance. To date, the scope of multiphoton investigations is limited by the restricted choice of excitation wavelengths provided by suitable laser illumination sources. In order to alleviate complex laser daily maintenance schedules and expand the spectrum of useful excitation wavelengths, new developments in femtosecond-duration optical pulse lasers are essential. Specifically, new laser sources must be engineered that are wavelength-tunable throughout the upper visible (500+ nanometers) and near-infrared portions of the spectrum.

During the past ten years, many new short-pulse laser system variations were made possible by the discovery of Kerr-lens mode locking in titanium-doped sapphire crystal lasers. Other new systems, such as diode-pumped solid-state and single-mode fiber femtosecond lasers are being investigated for their potential as excitation sources for multiphoton microscopy. In the future, frequency doubling of femtosecond pulsed lasers by optical parametric oscillators (**OPOs**) may provide a universal solution to cover the useful wavelength range. Present applications of lasers in microscopy are expanding rapidly in the areas of confocal microscopy, optical trapping, and the release of caged compounds and fluorophores. The development of compact solid-state lasers with emission lines in the blue-green and ultraviolet regions of the spectrum should serve to further increase the utilization of these devices in microscopy.

Chapter

8

The Application of Microscopes

In this chapter we will discuss about the application of different microscopes which has been explained in the previous chapters.

Familiarization: First, familiarize yourself with all the parts of a microscope so that you can easily move from one part to another during operation.

Moving and Placement: It is worth remembering that while a good quality microscope will last a life time, it is a sensitive scientific instrument that will suffer damage from sharp blows or impact. Always, therefore, carry your microscope in both hands. Grasp the arm with one hand and place the other hand under the base for support. Always place the microscope on a level and stable surface.

Slide Preparation: Microscope slides should always be prepared with a cover slip or cover glass over the specimen. This will help protect the objective lenses if they touch the slide. To hold the slide on the stage fasten it with the stage clips. You can push down on the back end of the stage clip to open it.

Focusing the Microscope

1. Start by turning the revolving nosepiece (turret) so that the lowest power objective lens is "clicked" into position. The lowest power objective is the shortest one. This objective is the easiest to focus and center the image in the field of view.
2. While looking at the objective lens and the stage from the side, turn the coarse focus knob so that the stage moves upward toward the objectives. Move it as far as it will go without touching the slide.
3. Now, look through the eyepiece(s) and adjust the illuminator and diaphragm until you attain the maximum, comfortable level of light.
4. Slowly turn the coarse adjustment so that the stage moves down (away from the slide). Continue until the image comes into broad focus. The turn the fine adjustment knob, as necessary, for perfect focus.
5. Move the microscope slide until the image is in the center of the field of view. Then readjust the illuminator or diaphragm in order to attain the clearest image.

6. Once you have attained a clear image, you should be able to change to a higher power objective lens with only minimal use of the focusing adjustment. If you cannot focus on your specimen, repeat the above steps and work from objective to objective until the higher power objective lens is in place.

General Advice

1. You should check to see if your microscope has a rack stop. If it does not have one, then be careful not to allow the objective lens to touch the slide as you may break the slide.
2. When using a monocular microscope, the correct technique is to look through the eyepiece with one eye and keep the other eye open. Most new users, tend to close one eye. While many microscopists do close one eye, you will help avoid eye strain by keeping both eyes open.
3. Finally, remember! When you view a specimen through a microscope, you are viewing an image through multiple lenses. As a result, the image is upside down and back-to-front so when you move the slide to the right, the image goes to the left and vice versa!

Care and Maintenance of Your Microscope

Your microscope will last a lifetime if cared for properly and we recommend that you observe the following basic steps:

1. When finished viewing , lower the stage, click the low power lens into position and remove the slide.
2. Switch off the microscope when not using.
3. Avoid touching the glass part of the lenses with your fingers. Use only special lens paper to clean the lenses.
4. Dust is the number one enemy of a microscope so always keep your microscope covered when not in use. When not in use for extended periods, replace the microscope in its box.

The ***stage*** is a platform that holds the slide containing the specimen to be viewed. A mechanical stage (see the photographs below) has a mechanism for moving the slide.

A light microscope must have a ***light source***. This is usually a light bulb located beneath the stage.

An adjustable ***diaphragm*** located beneath the stage is used to regulate the amount of light that passes through.

A ***condenser*** contains two sets of lenses that concentrate light. It is located directly underneath the stage. Light from the light source passes through the diaphragm and condenser before continuing up through the specimen to be viewed.

The ***body tube*** contains an ***ocular lens*** (eyepiece) and a ***nosepiece*** with several ***objective lenses***. Each objective lens is used for a different magnification and is moved into place by rotating the nosepiece. The image is brought into focus by adjusting the ***coarse*** and ***fine focus*** knobs.

Binocular microscopes have two eyepieces; monocular microscopes have one and the trinocular have an extra third eye piece for the camera. The distance between the two ocular lenses of a binocular microscope can be adjusted to fit the distance between your eyes.

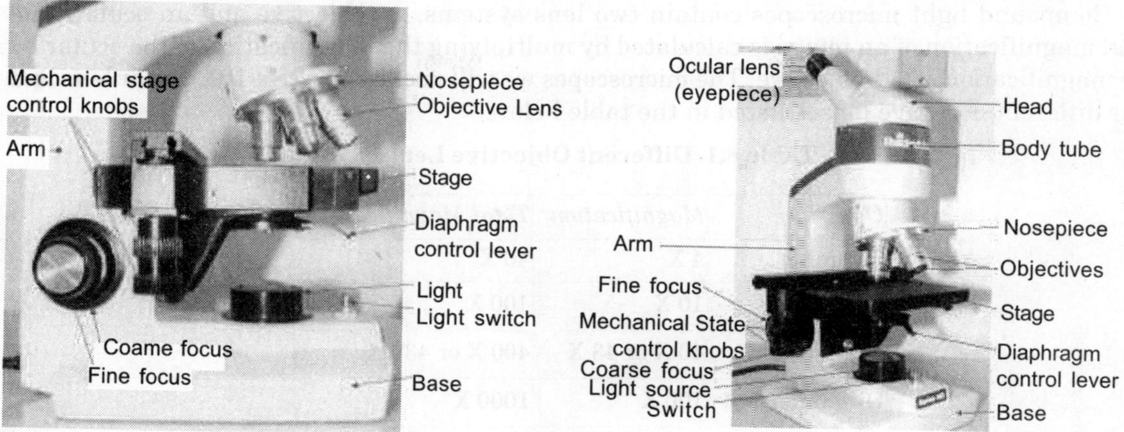

Fig. 8.1. Showing the Different Adjustments. **Fig. 8.2. Showing the different parts used.**

Other optical devices such as binocular telescopes and field glasses also have two ocular lenses that adjust in a manner similar to the microscope. Binocular lenses can be adjusted individually, making it unnecessary for many people to need their glasses when using them. If you wear glasses and are unfamiliar with adjusting binocular lenses to correct for your own eyes, see the section titled "Adjusting the Ocular Lenses".

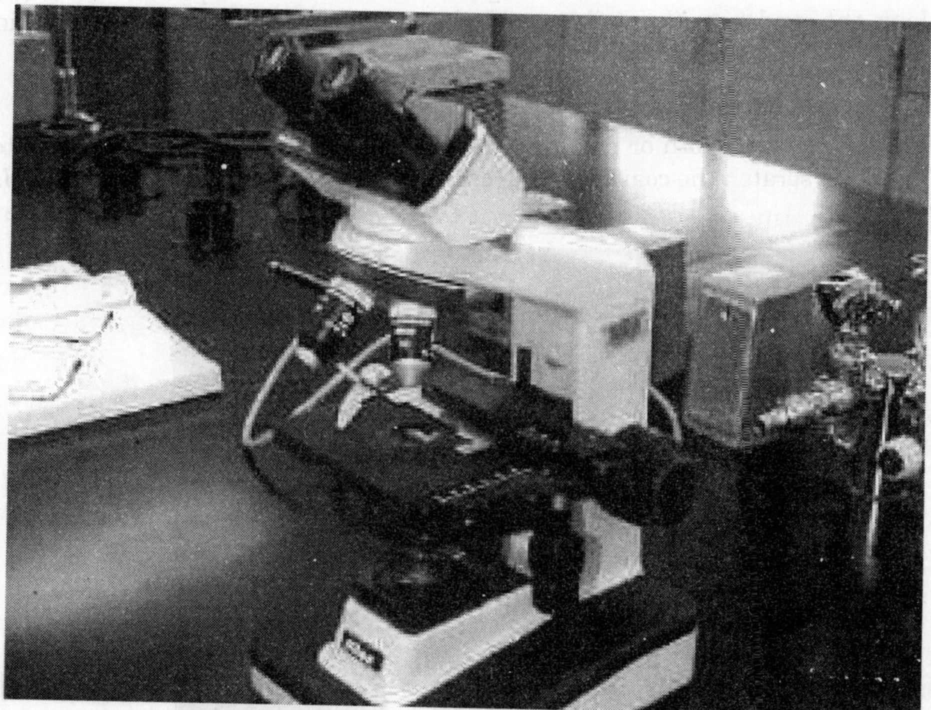

Fig. 8.3. The Phase Contrast Microscopy in the Application and Use

Theory

Compound light microscopes contain two lens systems, an objective and an ocular. The total magnification of an image is calculated by multiplying the magnification of the ocular by the magnification of the objective. The microscopes we will use each have a 10X ocular lens and four different objective lenses listed in the table below.

Table 8.1. Different Objective Lenses.

Objective	*Magnification*	*Total Magnification*
Scanning	4 X	40 X
Low Power	10 X	100 X
High Power	40 X or 43 X	400 X or 430 X
Oil Immersion	100 X	1000 X

Light bends when it passes from glass to air or from air to glass because air and glass have different refractive indices. The bending of light as it passes through the glass slide to the air and then to the glass lens decreases the resolving power. At high magnification (1000X) it can prevent a clear image from being viewed. This decrease in resolution can be prevented by putting ***immersion oil*** between the slide and the lens because immersion has the same refractive index as glass.

The condenser also increases the resolving power of the microscope. When using the oil-immersion lens, the condenser (located beneath the stage) should be raised to a position very close to the stage for maximum resolution.

Using the Microscope

CAUTION - Never use cloth or paper products (paper towels, tissue paper, etc.) to clean the lenses. They will scratch the coating and decrease the resolving power of the lens. Use only ***lens paper***.

Fig. 8.4.

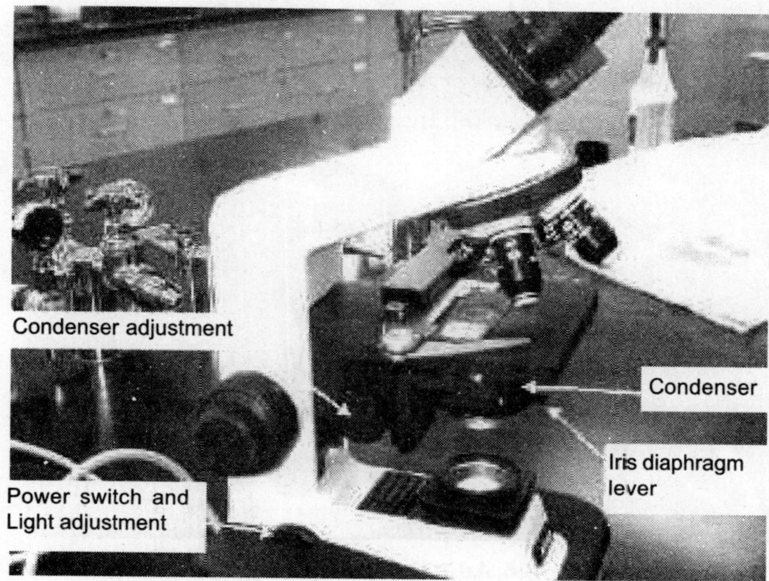

Fig. 8.5.

Switch the microscope to the lowest magnification or raise the objectives from the stage before inserting a slide. This will prevent the objective lens from being accidentally scratched by the slide.

Place the slide to be viewed on the stage and center the specimen over the opening.

Begin with either the scanning lens or the low power objective lens.

Raise the stage (or lower the lens) all the way so that the slide is as close as possible to the objective lens.

Use the coarse adjustment know to slowly raise the lens from the stage while viewing the image. Fine focusing is not needed when using the lowest magnification (scanning or 4X objective). If you are using any of the other objectives, it will be necessary to use the fine focus after using the coarse focus.

Adjust the condenser so that a sharp focus is produced. This step is important at the highest magnification (oil immersion or 1000X).

Adjust the iris diaphragm. This will need readjustment after changing to a different magnification.

To Increase the Magnification

The microscopes are ***parfocal***, meaning that after you adjust the focus, the image will remain approximately in focus if you change the magnification.

Center the object before switching to a higher power objective. This will help you find the object after switching the objective.

Switch to the next highest power. It will be necessary to center the image again. The image should be approximately in focus but it will be necessary to use the fine focus. The coarse focus should not be needed after switching objectives.

Fig. 8.6. Adjust the diaphragm.

This procedure is repeated each time you switch to a higher magnification.

Application and Using Oil Immersion

The 100 X objective (1,000X total magnification) requires that a drop of immersion oil be placed between the slide and the lens.

After focusing the specimen under high power (400X or 430X, see above), rotate the high power objective out of the way and place a drop of immersion oil on the slide. Rotate the oil immersion objective into place so that it touches the oil.

Adjust the fine focus, condenser, and iris diaphragm as previously described.

After viewing with oil, the lens must be cleaned with fluid designated for this purpose. Remember, use lens paper only. ***Never*** use cloth, paper towels, or other paper products on coated optics.

Practice with the Microscope

• Obtain a slide of colored threads and view them under the scanning and low power. Use the focusing procedure described above.

(1) Can you tell which thread is above the other?

View the threads under high power (400X or 430X). Use the fine focus to focus to determine the order of the threads from top to bottom. As you rotate the fine focus, different strands will go out of focus while others will become more sharply focused. This procedure will therefore enable you to determine the order of the threads.

(2) Are all of the threads in focus at the same time?

(3) What is the order (from top to bottom)?

"Depth of field" refers to the thickness of the plane of focus. With a large depth of field, all of the threads can be in focused at the same time. With a smaller or narrower depth of field, only one thread or a part of one thread can be focused, everything else will be out of focus. In order to view the other threads, you must focus downward to view the ones underneath and upward to view the ones that are above.

(4a) What happens to the depth of field when you increase to a higher magnification (increases, decreases, or remains the same)?

(4b) Explain how the slide with threads could be used to answer the question above.

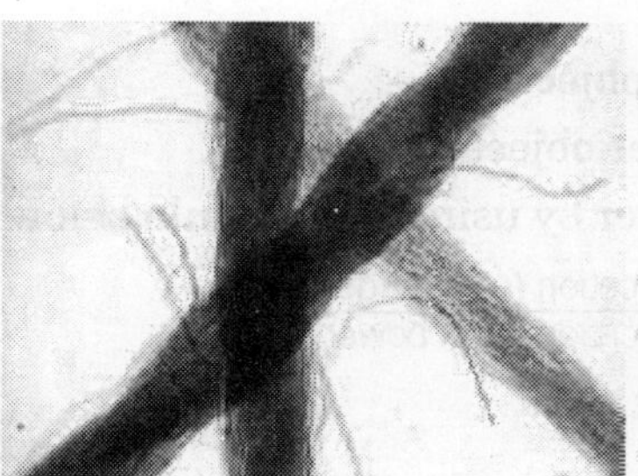

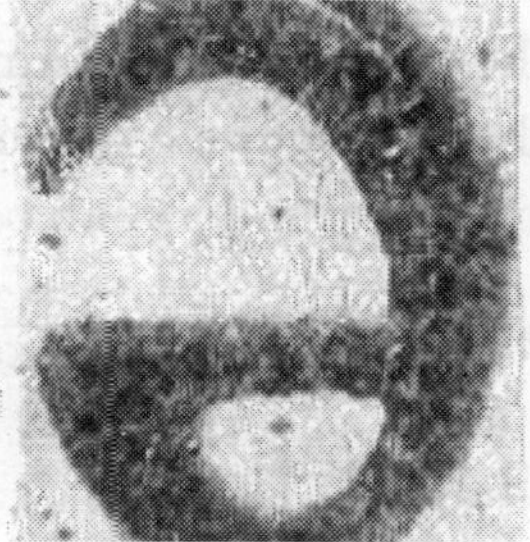

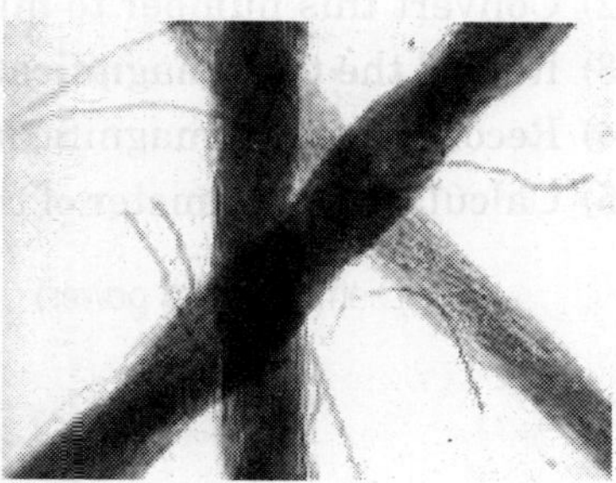

Fig. 8.7. A B and C The Different objects under Microsopes.

- To obtain a slide of the letter e and view it under the scanning objective. Move the slide to the left.

(1) What way did the image move when the slide was moved to the left?

(2) How does the orientation of the image compare to the image on the slide?

Paramecium

- Obtain a prepared slide of paramecium and view it under high power.

Draw the cell. Label the cilia and the oral groove.

Determining the Size of a Specimen

It is useful to know the size of the field of view so that you can estimate the size of objects. If the field of view is 2 mm and an object is about one half that big, then the object length is ½ × 2 mm = 1 mm.

We can measure the diameter of the field of view under low power using a ruler. You cannot use a ruler under high power because it is too big.

Example #1

Suppose you measure the low-power field of view with a ruler and it is 2 mm.

If high power is 10X more magnification than the low power, the field of view will be 1/10 as big. The field of view under high power will be 2 mm X 1/10 = 0.2 mm.

Example #2

Suppose that the low power diameter (LPD) is 2.5 mm (2500 um).

LPM (low power magnification) = 100X (10X objective and 10X eyepiece)

HPM = 400X (40X objective)

What is the HPD? In other words, if 100X is 2500um wide, how wide is 400X?

Answer: It is 1/4 as wide. = 2500 × 100/400

Your Microscope

• Place a small transparent ruler on the stage of your microscope and measure the diameter of the field of view using the scanning (4X) objective.

(1) Record the diameter in millimeters.

(2) Convert this number to micrometers.

(3) Record the total magnification when using the scanning objective.

(4) Record the total magnification when using the high power objective.

(5) Calculate the diameter of the field of view under high power by using the formula below:

$$\text{Diameter (high power)} = \text{Diameter (scanning)} \times \frac{\text{Magnification (scanning)}}{\text{Magnification (high power)}}$$

$$= \#2 \text{ (above)} \times \frac{\#3}{\#4}$$

Making Wet Mounts

Wet mounts are useful for viewing living biological material.

To make a wet mount, place the specimen on a slide and then add a drop of water or stain. Stain is often used to make the specimen more visible.

Place a cover slip at an angle so that it touches the drop. Slowly lower the raised end of the cover slip. The diagram below shows that as the cover slip is lowered, the drop of liquid moves to the right (Fig. 8.8).

Cheek Cells

• Scrape the inside of your cheek with a toothpick and rub it on a dry slide.

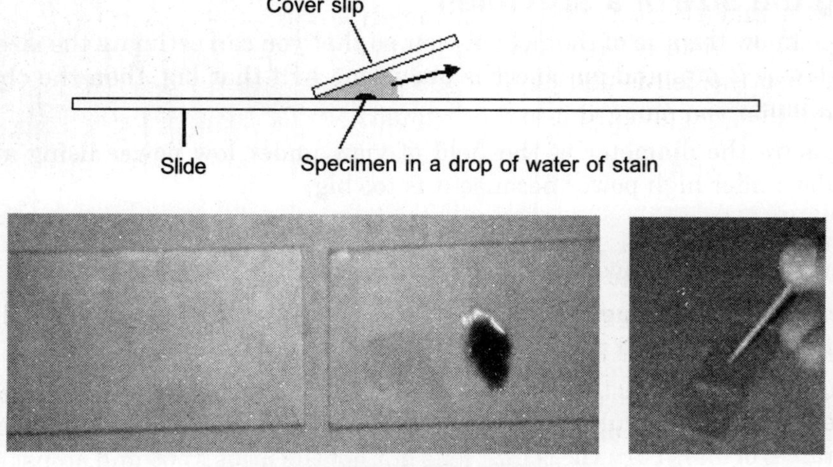

Fig. 8.9. The slide Preparation

- Add one drop of methylene blue to stain the cells. This will make them easier to see.
- Place a cover slip on the slide as described above and observe the cells under low power then high power.

  (1) Draw a cell under high power.

Below: Cheek cells 100X and 400X.

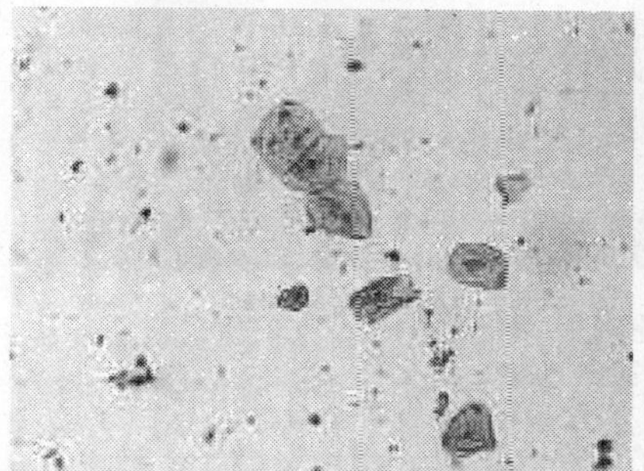

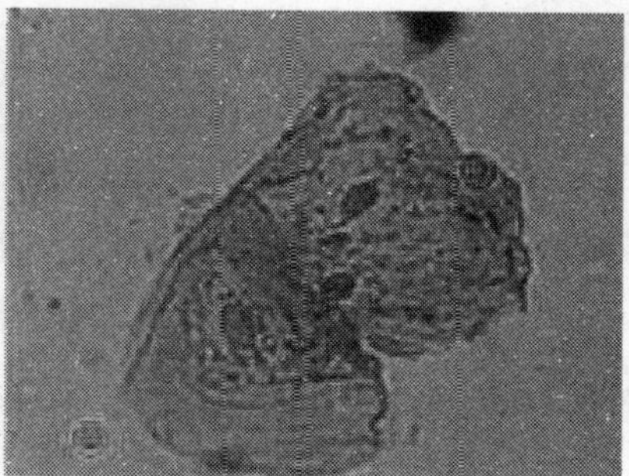

Fig. 8.10. The Different objects under Microsopes

Size of Onion Cells

- Prepare a wet mount of fresh onion and stain it with methylene blue. Try to get the thinnest piece that you can; it should be thinner than paper. It might help if you take a thicker piece and bend it until it snaps into two pieces. After it snaps, there is usually a very thin connection remaining between the two pieces. This thin connecting layer can be peeled off and placed on a microscope slide for viewing. After you view this slide and answer the questions below, set the slide aside for use later in the laboratory period.

2) Draw several onion cells as they appear under high power.

3) Estimate the length of a typical onion cell in micrometers (um). This can be done by using the diameter of the field under high power calculated earlier. The following is an example; your numbers will probably be different. Suppose that a typical cell is approximately 1/3 the diameter of the field of view and the field of view is 450 μm, then the cell is 450 μm/3 = 150 μm. Show the numbers that you plugged into the formula.

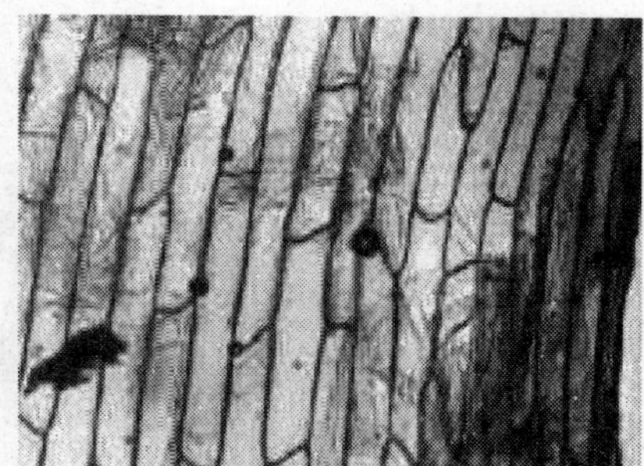

Fig. 8.11. The Onion Cells under microscope.

Adjusting the Ocular Lenses

Binocular microscopes and stereomicroscopes have one ocular lens that is adjustable. This enables you to adjust the viewing for your eyes so that you do not need to wear your glasses. Remove your glasses before doing the procedure.

To adjust these lenses, first, cover the adjustable lens (or your eye) so that you cannot see the image through it. Next, focus the microscope the way you normally would so that a sharp image is produced through the ocular lens. Cover this lens (or eye) and view the image through the other (adjustable) ocular. Turn the ocular lens but not the focus knob and adjust so that the image is sharp.

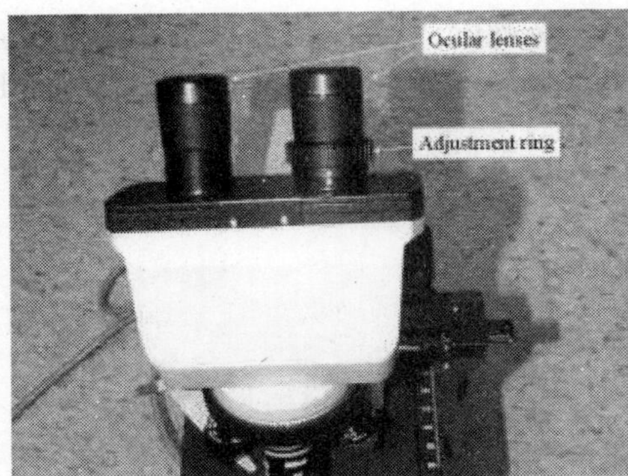

Fig. 8.12. The Binocular and trinocular Heads and Objectives.

Binocular Microscope

Binocular dissecting microscopes are useful for viewing material that is too large to be viewed by compound light microscopes. The magnification of these microscopes typically ranges from 8X to 40X.

Dissecting microscopes (Stereomicroscopes) have two ocular lenses and produce a three-dimensional image. If you wear glasses and are unfamiliar with adjusting the ocular lenses of optical equipment to suit your eyes.

There are several kinds of binocular dissecting microscopes available in the laboratory room. The microscope shown below requires an external light source and the magnification is changed by rotating the large knob. These scopes have a 15X ocular lens and an additional 2X lens screwed onto the objective. The total magnification is 20X, 25X, 40X, 60X, or 80X. If the 2X lens is removed, the magnifications become 10X, 15X, 20X, 30X, and 40X. The total magnification of these microscopes can be read directly from the magnification adjustment knob without doing multiplication.

The microscope below requires a separate light source (not shown).

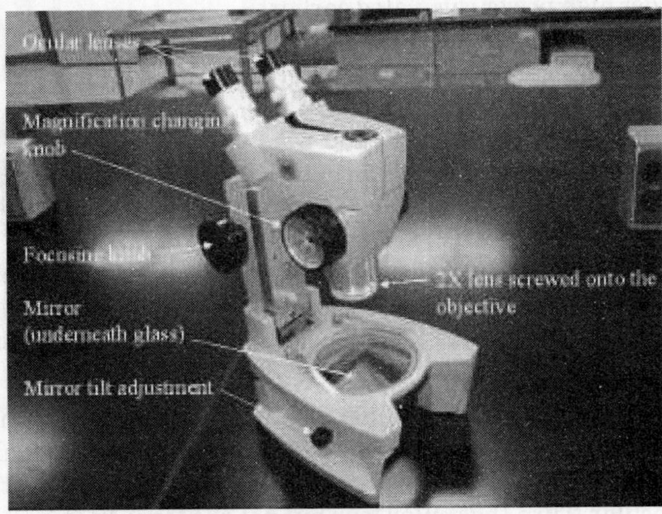

Fig. 8.13.

The microscope shown below (Fig. 8.14) contains a zoom adjustment. By rotating the adjustment, the magnification changes from 8X to 40X.

The microscope requires a separate light source.

Both microscopes have zoom magnification adjustments. The scope on the left ranges from 20X to 40X and the one on the right ranges from 10X to 45X.

These microscopes have a light source built in. The light adjustment knob enables the specimen to be illuminated from above, from below or from both above and below at the same time. The zoom adjustment of the microscope on the right has the total magnification written on it. The zoom adjustment on the microscope on the left has numbers that must be multiplied by 10 to give the total magnification.

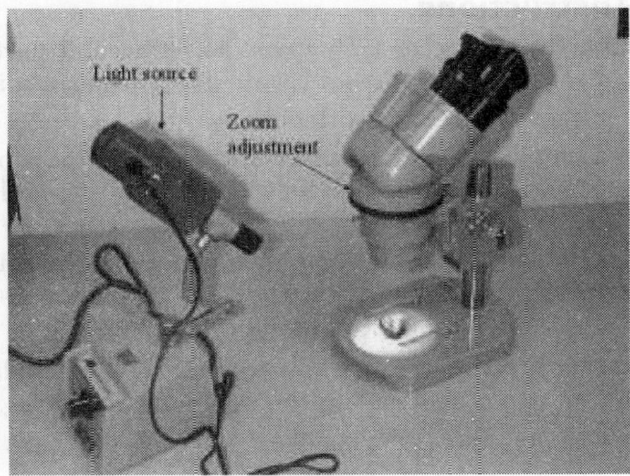

Fig. 8.14.

Practice Using the Dissecting Microscope

- Obtain a dissecting microscope and a separate microscope lamp.

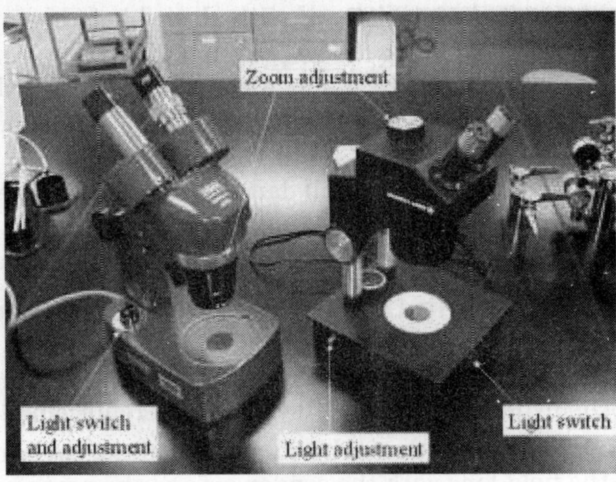

Fig. 8.15.

View a plant leaf using the dissecting microscope by placing the entire leaf under the scope. It is not necessary to prepare a slide.

The leaf will probably look best if illumination is provided from below.

• View a clam shell using the dissecting microscope. The shell is opaque, so it must be illuminated from above.

Using Phase Contrast Microscope

[Note: you are encouraged to print these instructions and to place them into your lab notebook as a handy reference anytime that you need to use the phase contrast microscopes.]

Using the Phase Contrast Microscopes

Common Sense Instructions

- Never Disassemble the Microscope. There are no serviceable parts inside the microscope. Disassembly will void the warranty and could degrade performance.
- Only plug in to standard 110/120V outlets.
- Use the Specified Lamp (Halogen lamp 6V-20W (Philips 7388) or halogen lamp 6V-30W (Philips 5761) along with the specified fuse 250 V, 1 amp, time-lag high-breaking, 5X20 miniature fuse X2.
- Turn off power switch before assembling the microscope, replacing the lamp or fuse and plugging in or unplugging the power cord.
- Keep the microscope free of moisture and foreign matter.
- The light becomes very hot during use. Do not remove the field lens unit while the lamp is on. Wait 30 minutes before touching the field lens unit after the lamp has been on.
- The rack for the stage protrudes during use. Be careful not to hit the rack when handling the microscope.
- Carry the microscope properly. One hand should be firmly grasping the upper rear base unit and the other hand should be underneath the lower base unit.
- Do not touch the glass part of the lamp with bare hands. Wear gloves or use a cloth when handling the lamp so that fingerprints are not left on the surface. Wipe off any fingerprints with a soft cloth moistened with alcohol before using the microscope.
- When changing specimens using the refocusing mechanism, gently lower the stage by hand taking care not to hit the field lens with the condenser holder.
- Do not turn the right and left focus knobs simultaneously in opposite directions.
- Do not turn the coarse focus knob any further after the stage has been moved up or down to its limit. These operations will damage the focusing mechanism.
- Use only a minimum quantity of oil for oil immersion. Surplus oil could flow onto the stage and then onto the condenser leading to degraded performance.

Replacing a Specimen Using the Refocusing Mechanism

Focus on a specimen with the 40X or 100X objective. Note that the specimen is very close to the objective. It is difficult to change the specimen without moving the focus knob. In a case like this, use the refocusing mechanism for easy specimen replacement.

- Use one hand to gently press down the stage. (When the lowering the stage take great care not to hit the field lens with the condenser and the parts under the condenser.)

- While holding the stage in that position, change the specimen.
- Gradually release the stage so that it rises slowly. The stage will return to the focal position.
- Always view large objects in the lowest magnification.

Using Oil-Immersion

The "Oil" mark on the side of an objective indicates that it is an oil-immersion type objective. An oil-immersion objective is used with immersion oil applied between the objective and the coverslip. For an oil-immersion objective with a numerical aperture of 1.0 or more, use of an oil-immersion type condenser is required to take full advantage of its performance. An oil-immersion type condenser, like an oil-immersion objective, needs immersion oil to be applied between the front of the condenser and the coverslip. The abbe condenser included in the bright viewfield set can be used for oil-immersion observation. The condenser has an oil receptacle around its front lens.

- **Handling Immersion Oil:** Use a minimum quantity of oil. Oil on the condenser could degrade optical performance. After completing oil-immersion observations, be sure to clean the objective, condenser and any other parts that may have been exposed to oil. Use a clean cotton swab dampened with absolute ethanol to remove the oil on the objective. Repeat with a second clean cotton swab dampened with absolute ethanol.
- **Application to the Condenser:** Move the specimen toward the back and lower the condenser slightly. Add a drop of oil on the front of the condenser through the long hole in the stage. Bring the specimen back over the condenser and slowly raise the condenser.
- **Application to the Objective:** Rotate the revolving nosepiece to move the objective out of position. Add a drop of oil to the specimen. Slowly rotate the revolving nosepiece to bring the objective back into position.
- **Eliminate Air Bubbles:** Air bubbles trapped in the oil between the coverslip and the objective lens degrade the image. If the image is poor and you suspect air bubbles try rotating the revolving nosepiece back and forth. If this does not work, add another drop of oil.

Parts of the Phase Contrast Microscope

See the parts of the phase contrast microscope in Fig. 8.16.

Using the Phase Contrast Microscope

- Turn on the microscope.
- Rotate the condenser turret so that the "A (hollow)" indication faces the front.
- Place the specimen on the stage and focus on the specimen with the 4X and then 10X Phase objective. (See above for details)
- Adjust the diopter and interpupillary distance. (See above for details)
- Center and focus the condenser. THESE ADJUSTMENTS ARE VERY IMPORTANT IN PHASE MICROSCOPY. (See above for details)
- Rotate the turret so that the "Ph1" indication faces the front.
- Place the green interference filter (GIF) on the filter holder around the field lens.

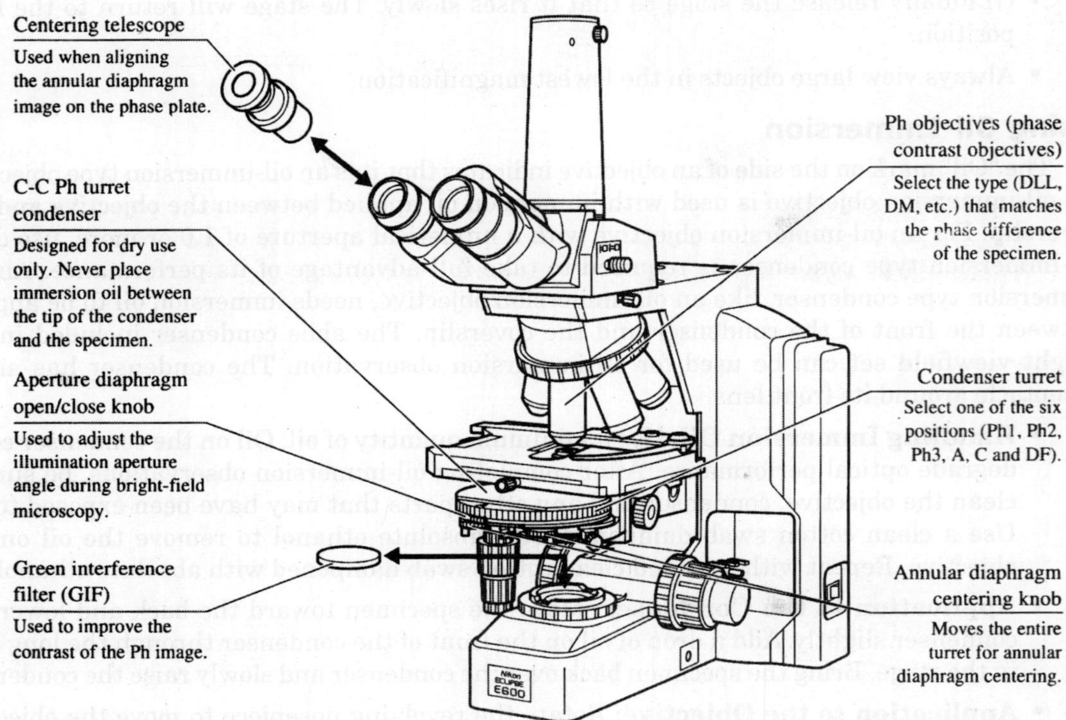

Fig. 8.16.

- Centre the phase annular diaphragm. THIS ADJUSTMENT IS VERY IMPORTANT IN PHASE MICROSCOPY.
- Adjust the field diaphragm so that it is just inside or outside of the view field.
- If you switch the objective, also switch the condenser turret so that it matches the phase code of the objective. At this time, check that the phase plate of the objective and the image of the phase annular diaphragm are aligned. Also readjust the size of the field diaphragm.
- If using an oil immersion type objective, apply immersion oil between the specimen and the objective.

Using the Bright-Field Microscopy

Perform the following for bright-field microscopy.

- Rotate the condenser turret so that the indication "A (hollow)" faces the front.
- Objectives with the magnification higher than 4X can be used for bright-field microscopy. Vignetting occurs on the view field periphery when using the 4X objective.
- The condenser performs just like an Abbe condenser described in the manual provided with the microscope.

Using the Dark-Field Microscopy

Perform the following for dark-field microscopy. Dark-field microscopy with the C-C phase turret condenser can be used as a finder to search for the object to be viewed in a specimen.

- Rotate the condenser turret so that the indication "DF" faces the front.
- Insert an objective having a magnification of at least 10X and a numerical aperture of no more than 0.7 into the optical path.
- Raise the condenser as far as it will go when viewing specimens with a thick glass slide. If the glass slide is too thick, the illumination may not reach the specimen surface. When this happens, prepare the specimen again using a thinner glass slide.

BIOLOGICAL STAINS

STAIN	*USE*	*SPECIMEN COLOUR*
Acid Fuchsin	Stains collagen fibers red and smooth muscle in contrast to collagen.	Red/magenta/violet
Congo Red	Used to stain amyloid. (A starchlike substance. A hard, waxy deposit consisting of protein and polysaccharides that results from the degeneration of tissue.)	Red
Crystal Violet	The (positive) dye of choice in Grams stain. Bacteria	Blue violet
Eosin B	Interchangeable with Eosin Y. Ionization – acid (Colour is red)	Bluish cast
Eosin Y	Stains alkaline cell parts (like cytoplasm). Use on plants, animals and blood.	Pink
Eosin Y ws	Common counterstain to alum hematoxylin and Eosin method.	Red
Gram Stain	Positive – Staphylococcus, Streptococcus, Bacillus subtilis (crystal violet) Negative – Neisseria, *E. coli* (Eosin Y ws)	Dark blue or purple Red
Hematoxylin	Stains nuclear chromatin using aluminum mordant. Acid resistant nuclear staining, muscle striations and some glial fibers (The delicate network of branched cells and fibers that supports the tissue of the central nervous system) with ferric salts mordant.	Yellow/brown
Iodine	Carbohydrates in plant and animal specimens Glycogen (Animal starch.. A polysaccharide, $(C_6H_{10}O_5)n$, that is the main form of carbohydrate storage in animals.	Brown or Blue-Black Red
Mercurochrome	Marks tissue margins and faces for orientation (Antiseptic). Similar to Eosin Y ws.	Red
Methyl blue	Stains collagen and connective tissue.	Blue
Methylene blue	Acidic cell parts (Use on animal, bacteria and blood specimens).	Blue
Saffron	Colours connective tissue yellow in contrast to the pink cytoplasm given by phyloxine.	Yellow orange
Safranin O	Commonly used for counterstaining nuclei red. Cartilage is stained yellowish.	Red
Toluidene blue	Stains acidic cell parts (like nucleus). Good to show mitosis in plant cells. MITOSIS = The process in cell division by which the nucleus divides, typically consisting of four stages, prophase, metaphase, anaphase, and telophase, and normally resulting in two new nuclei, each of which contains a complete copy of the parental chromosomes. Also called karyokinesis.	Dark Blue